FOSSILS

RICHARD FORTEY

Published by the Natural History Museum, London

First published by the Natural History Museum,
Cromwell Road, London SW7 5BD
© Natural History Museum, London, 2009

The first edition of this book was published in 1982.
This edition has been completely revised and updated.

ISBN 13 9780 565 09233 7

A catalogue record for this book is available from the British Library.

Designed by Mercer Design
Reproduction and printing by Craft Print, Singapore

Front cover: *Erbenochile* trilobite found in Anti-Atlas mountain range,
Morocco © NHMPL
Front cover flap: Jurassic ammonite, *Promicroceras planicosta*, Somerset,
England © NHMPL
Back cover: Eocene bony fish, *Pristigenys substriatus*, Monte Bolca, Verona,
Italy © NHMPL
Page 4: A section through the nautilus, *Nautilus pompilious* © NHMPL

Contents

Preface

I wrote the original edition of *Fossils: The Key to the Past* more than 25 years ago. It is a measure of the vigour of palaeontology as a science that many new discoveries have meant that I have had to repeatedly revise parts of the text to bring it up to date. This is true particularly of this latest edition into which new chapters have been introduced to take in some of the recent discoveries about mankind's history, and to nod briefly towards the new science of molecular palaeontology. The illustrations have been revisited, and additional ones have been added.

Like all science, palaeontology moves forward or it dies. Nonetheless there is still a place for a straightforward introduction to the study and meaning of fossils: how they help us understand the geological past, how their fascination is always more than just 'stamp collecting'. Even though new discoveries continually add to our comprehension, the fundamentals of the science do not change so radically. I hope that I have retained the virtues of an accessible guide to what palaeontology does, and why it still matters.

This edition will appear one hundred and fifty years after the publication of *The Origin of Species by Means of Natural Selection* by Charles Darwin. Many of the questions that puzzled Darwin about the fossil record have now been answered. Links between major groups of organisms have been discovered – like the relationship of birds to dinosaurs. What were 'holes' in the record to Darwin, such as the Precambrian record of life, have now been largely filled in. But unsolved problems still remain, and that is how it should be – it would be a poor science that had all the answers.

The purpose of this fourth edition of *Fossils: the key to the past* remains the same as when it was first published: to stimulate the reader to further study and enjoyment of our geological legacy.

RICHARD FORTEY, JANUARY 2009

Acknowledgements

Special thanks to Chris Stringer for helping me update the hominid section, and reading that part of the book, and to Brian Rosen for supplying photographs from his personal collection. Thanks also to my colleagues who looked out new pictures, and who checked caption text. Thanks, too, to the Publishing team for handling the logistics and proofing of this edition after my formal retirement from the Museum.

Buried in the rocks

Nearly everyone has seen fossils at one time or another: dinosaurs in the Museum displays, ammonites collected from a seaside holiday, or odd-shaped stones picked up on a country walk. The science of fossils is palaeontology, which is a long word for the business of studying the life of the past.

RIGHT: Scientists at a dinosaur excavation in Niger, 1988, cleaning the exposed elements of the fore and hind limbs of a sauropod dinosaur.

This may create an image of a palaeontologist as a fusty old professor, as desiccated as some of the fossils he studies, busying himself in drawers thick with dust. But, like most other sciences, palaeontology has been growing in the last decades, and there have been many new and exciting discoveries and new ways of looking at 'old' fossils. The study of fossils is inextricably linked with the study of the rocks that contain them, and the fossils provide some of the crucial evidence for the history of the last 1,000 million years or more. When one thinks of all that has happened in the last two *thousand* years, to be a historian of a thousand million seems an impossible task. Palaeontologists know that there will always be new fossils to discover. This book will try to show some of the things that can be done with fossils, as well as describing the fossils themselves. It will show how these dead objects can be brought back to life, how they are classified and how they can be used to read the details of past climates and the changing distribution of continents. So however mundane a small fossil held in the palm of your hand may seem, it can be the key to unlocking some of the most profound secrets of the Earth.

Fossils are the remains of prehistoric animals or plants. Usually they are some hard part of an extinct organism, resistant to decay, that has been preserved enclosed in sediment — past life that has been buried in the rocks and entombed inside them. Fossils could equally be the rock record of activity of animals: fossil footprints, perhaps, or the tubes and trails of soft-bodied worms that otherwise leave no trace.

The province of the palaeontologist overlaps with that of the archaeologist, but generally palaeontologists concern themselves with the organic remains of greater antiquity than the recorded history of humankind. A cow that died 500 years ago and is then exhumed as a skeleton is in a sense a fossil, but its relevance is towards an understanding of the agricultural practices of the time. Archaeology is concerned with the history of organised humans over the last 5,000 years or so. Most of the fossils we shall be discussing in this book are many times older than this, but of course in the sites where the remains of early humans are discovered it is quite possible to find the palaeontologist and the archaeologist working side by side.

BELOW: An Early Cretaceous fossil lizard, *Adriosaurus suessi*, from the Isle of Lesina, Dalmatia, Croatia.

2. Fossils are often small and insignificant, but they can be spectacular, and it is likely that they attracted the curiosity of humans from the earliest times. They have been found as 'charms' in the caves occupied by our remote ancestors. It is a remarkable fact that their significance has been appreciated for less than two centuries. Compared with physics or mathematics, palaeontology was a science that had a very late development.

Ancient Greek philosophers made observations on the formation of rocks that seem with hindsight to have been remarkably perceptive, but these early investigations do not appear to have excited the curiosity of the Renaissance scientists. Always an acute observer, Leonardo da Vinci noted the presence of fossil shells in rocks now far removed from the sea and speculated that the rocks must have formed beneath water. His opinions lay largely unheeded, buried in his notebooks. Perhaps the truth is that science had more pressing problems to solve: the nature of motion; the structure of the solar system; the reality of the elements. Until the 18th century, observations on fossils were principally the province of the antiquarian who would illustrate such 'figured stones' along with other curiosities of the natural world. Nonetheless, collections were being made and housed in universities like Oxford and Cambridge, and these provided the raw material for subsequent interpretations.

The attention focused upon fossils was part of the great surge in interest in the ordering of the *living* organisms of the world. Early accounts of living animals and plants had often been fanciful and inaccurate, and there had been little attempt at a rational ordering based upon their shared similarities. During the last part of the 17th and 18th centuries all this was to change: it was an age that felt conviction in the presence of order in the world. At the same time the standard of illustration and description of plants and animals improved enormously, partly because there was a market for well-produced books on nature in the libraries of the rich. The work of John Ray (1627–1705) in England, Georges Buffon (1707–1788) in France and Carl von Linné (or Linnaeus) in Sweden (1707–1778) set out the outline of classification of organisms still in use today. Of pre-eminent importance was the recognition of *species*, the different kinds of animals and plants, which formed the units of classification. Linnaeus

ABOVE: A 70 million year old *Edmontosaurus regalis* skeleton still half buried in sandstone rock, from the Upper Cretaceous of Alberta, Canada.

evolved the hierarchical classification of organisms that we still employ, and to a certain extent naming something is a step towards understanding it; it gives a common vocabulary, so that scientists the world over know that they are talking about the same plant or animal when they use its scientific name.

It was a natural extension to apply the Linnaean system for naming organisms to their fossil equivalents; the framework for their logical arrangement had been established. Linnaeus himself paid little attention to the sort of invertebrate animals that constitute the bulk of common fossils, but Jean Baptiste Lamarck (1744–1829) made a special study of such humble organisms and was able to apply his researches on living animals to the naming of fossil forms. Up to this time it was a relatively open question whether species were 'fixed' or not and Lamarck firmly believed that one species could change to another by the inheritance of advantageous characteristics. But belief in the fixity of species became more generally accepted, for it was espoused among others by the great comparative anatomist Baron Cuvier (1769–1832), and seemed to be consistent with a religious view requiring the special creating hand of God in organising the wonderful diversity of the natural world, past and present alike. "God made them high and lowly, he ordered their estate", as the hymn expresses it.

THE GEOLOGICAL BREAKTHROUGH

The science of geology developed as a respectable discipline at about the same time, a product of a similar urge to observe directly from nature and synthesise the observations into principles. Theories about the significance of fossils could not be divorced from ideas about the formation of the rocks that contained them. As so often happens, the early ideas had a compelling simplicity: in the 18th century the influential mineralogist Abraham Werner thought all rocks were accounted for by a great event of precipitation from a universal ocean. The coarse-grained rocks (which we now recognise as igneous in origin) were deposited first, followed by sedimentation of all the overlying rocks, resulting in the layering of sedimentary formations. Such an event required little time and was consistent with a biblical notion of Creation. The idea of vast stretches of 'geological time', which now seems familiar enough for even small children to talk glibly of millions of years, would have seemed quite alien then. The origins of our ideas in Britain go back to one of the great classics of geology, *The Theory of the Earth*, published by a Scot, James Hutton (1726–1797) in the closing years of the 18th century. Hutton based his book on detailed field observations of rocks, and recognised the importance of looking at the formation of present-day sediments in the interpretation of their ancient equivalents. There was nothing particularly mysterious or catastrophic about the past recorded in the rocks; to understand the past it was only necessary to look at the processes at work in the world today. But of course the formation of all these rocks required huge expanses of time, and creation must have been a protracted process. Hutton's ideas took root, but only slowly. In the earlier part of the 19th century the most widely accepted beliefs on the formation of rocks were a more sophisticated development from Werner's: the varied faunas and formations were the product of successive catastrophes, after which the world had been repopulated by new animals and plants, created afresh, to set the stage for the creation of the present world as explained in biblical revelation. The last catastrophe, and not even the most dramatic, was the Flood. Evidence of the Flood could be found in the cave deposits, which were the youngest known (what we would now call Pleistocene), where the remains of bears,

ABOVE: A portrait of James Hutton by Henry Raeburn. Hutton was a Scottish scientist and geologist who wrote *The Theory of the Earth* in 1795, which introduced the idea that rocks formed over vast periods of time.

rhinos and elephants could be found in areas where they do not live today. William Buckland's *Reliquiae Diluvianae* (1824) documented many of these in impressive detail.

The first half of the 19th century was also a time of great practical enterprise, when pure science and technology continually interacted to throw up new inventions of immediate relevance to a burgeoning industrial society. The discoveries of pragmatic men without formal scientific training could proceed independently of the current scientific controversies on theories of the Earth. When William Smith (1769–1839) published his geological map

LEFT: William Smith's geological map was published in 1815. He used the fossils he had found in different rocks as signatures to map the strata of different rock types throughout England and Wales.

of England and Wales in 1815 its practical importance was obvious. By mapping out the formations of rocks you could tell the best route through which to push a canal, or where to obtain clay for brick-making. Smith had recognised the importance of fossils in characterising the various strata he coloured on his map. He had collected these fossils from the trenches he cut for his canals, and from small quarries and pits cut for lime or bricks, that were scattered over the English countryside. The fossils varied in kind according to the position from which they were recovered: older rocks evidently lay progressively beneath the strata that covered them. And so, piece by piece, an atlas of fossils could be compiled which identified the strata and the order in which they occurred. Once the pattern was known, a traverse into strange ground could be interpreted simply by examining the fossils culled from the local exposures. The zoological relationships of the fossils did not matter too much (in the same way that you do not have to know the name of a criminal in order to be able to identify him from his fingerprints). Fossils were a practical and effective method of solving the problems of the structure and order of the rocks in England. When he was presented with the Wollaston Medal of the Geological Society in 1831, Smith was dubbed 'The Father of English Geology' by the President, a tag that has remained. Similar elucidation of rock succession and the fossils enclosed within them was being pursued in France, and by the 1820s it was clear that fossils were not merely objects to divert the antiquary, but were of

BELOW: Fossil remains of a hand that once belonged to the herbivorous dinosaur, *Iguanodon*. It was was one of the first dinosaurs to be described together with *Megalosaurus* and *Hylaeosaurus*.

great practical importance in solving geological problems, and many of these in turn were of direct economic relevance. Fossils have continued in this role ever since, and in a sense the palaeontologists employed by oil companies today to identify fossils owe their jobs to William Smith and his contemporaries.

In the 20 years from the publication of Smith's map, the government too had begun to realise the implications of the kind of geological mapping and collecting that Smith had pioneered, and the Geological Survey of Great Britain was accordingly founded. The early officers of the Survey include some of the most distinguished names in geology. Their successors are far more numerous, for more knowledge always results in new techniques for answering yet more questions. And the more searching that was done, the more new fossils came to light, many of them stranger and more exciting than could have been dreamed of 50 years before. In 1825 the description of the first dinosaur bones was published. So much was discovered that scholars became specialists in the description and identification of particular groups of fossil animals and plants. Money was available to publish great compendia of such descriptions, and most of the fossils were described therein for the first time. In the great age of the amateur naturalist there was a good deal of public interest in such monographs, and local natural history societies seemed to spring up wherever there were enough people to form a committee. The study of rocks became respectable employment for gentlemen of private means. The period from the 1850s to the 1890s was in many ways the heyday of palaeontology, for books about fossils were only exceeded by the discovery of new kinds of extinct organisms. Many of the classics of the subject were written then, and the fossil faunas of Europe and North America were gradually exposed to public view in museums. In the process it slowly became commonplace to think of larger and larger stretches of geological time during which the strange former inhabitants of the Earth reached their prime, and perished, to be replaced by others.

Geological mapping proceeded over a wider and wider area, and time and again fossils proved their worth in determining the order of succession and identifying the subsequent events that twisted and distorted the rocks into the structures they have today. Such studies changed the way the landscape itself was observed: this high range of hills was the product of a convulsion of the Earth which folded the rocks into complicated piles; that valley was the floor of an ancient lake, the water from which had long since drained into the sea. As knowledge increased, so the explanation of the shaping of the Earth by a series of catastrophes came to seem excessively arbitrary. James Hutton's ideas of interpreting the past by observation of present-day processes were more persuasive, and received cogent and sweeping support from the works of Charles Lyell, especially his *Principles of Geology* (1830–1833). The past became subject to the same natural laws that prevail today, intensified in their operation at some times to be sure, but the mysteries of the rocks could be unravelled by observations from volcanoes, seas and winds that can be directly recorded.

The validity of this way of looking at things has remained; despite the realisation that the most distant past of the Earth was different from the recent in several important ways, geologists still look at the formation of present-day sediments to interpret the features of ancient rocks. Those who accepted Lyell's book had also to accept another idea. It was common observation that rocks, which had originally been deposited beneath the sea, now formed huge cliffs, and when the thickness of *all* such sedimentary rocks from the different periods of geological time were piled on top of one another the total must be immense.

Yet sediments accumulate slowly, a centimetre or two a year. So it must have taken an inconceivably long time to accumulate this great pile of sedimentary rocks; geological time must be reckoned in millions, not thousands of years. The biblical Creation story could not be literally true. And since fossils of many kinds were found in the sedimentary rocks, life must have been present on Earth for a comparable period of time, time enough for the changes between one kind of animal and another to have happened. It was not necessary to have catastrophic extinctions followed by re-creation of new life forms. The living biological world could have originated by a process of transformation, just as the world itself was slowly shaped by the same forces that operate in it today. The marriage of geological time with transformations of one species of animal or plant into another led to one of the simplest, but one of the most important and influential ideas: the theory of evolution.

Charles Darwin published his great book *On the Origin of Species by means of Natural Selection* in 1859, almost 30 years after Lyell's *Principles of Geology*. The title would scarcely recommend itself today as a bestseller, but that is what it was, sold out as soon as published. The *idea* of evolution was not in itself a new one; it can be found implicitly or explicitly in many earlier works, notably those by Lamarck. The time was propitious for a summary of the kind that Darwin provided, and he described a mechanism that drove the process of change. Phrases that originated (or were garbled from) Darwin's book have entered the language – "the survival of the fittest", "Nature red in tooth and claw" (coined by Darwin's champion T.H. Huxley). Perhaps it was this somewhat stark view of nature, not Mother Nature at all, that was partly responsible for the furore that was caused by Darwin's book. There was also the clear implication that man, as another animal, was subject to the same inexorable process. If Darwin was correct the boundary between man, with a soul, and the ape without one was a slim one indeed. The orthodox clergy became ranged on one side, and many, but by no means all, of the leading scientists on the other, and it was a number of years before the concept of evolution became generally acceptable. There are still religious fundamentalists who regard the battle as continuing. In fact, Darwin shared the first proposition of his theory with another biologist, Alfred Russel Wallace, who had independently reached similar conclusions, although Darwin's book indicated the broader scope of his ruminations on the subject. But this does show that the time was ripe for such a change of outlook on the origins of the diversity of life. That it was Darwin who expressed the new view was certainly no coincidence. By the time that *Origin* was published, he had already written several papers that were (and have remained) classics of their kind, notably on the origin of coral reefs. He was a consummate observer.

ABOVE: When Darwin published *On the Origin of Species by means of Natural Selection* in 1859, the idea of evolution had already been under discussion, but what Darwin provided was a mechanism for evolution. Altogether he wrote sixteen books and there are many editions of the titles. All his writings remain in publication today in various forms and languages.

Many of his ideas probably had their origin in the world voyage he made on the *Beagle* (1831–1836), and the genesis of many of them are to be found in his diary of the journey, and in his letters. Even after the publication of his most famous book, which would have been enough for most biologists, he continued to make fundamental contributions to our way of looking at the natural world, all of them founded on meticulous observations, in some cases extending over decades. He was the quintessential biologist.

The combination of Darwin's ideas with the expanded geological timescale gave a new impetus to palaeontology. The rocks recorded the very history of evolution. Darwin himself recognised the imperfection of the fossil record, and devoted some space in the *Origin of Species* to explain why this was so, but the theory provided a new framework in which to interpret fossil remains. Argument still continues over just how much the fossil record can reveal, but the fact is that diligent search of the rocks has uncovered animals that show how one group links up with another. Inevitably the search extended for the ancestors and relatives of humans themselves, and eventually candidates for this role were discovered in Africa, although it took a century to find them. Advances in other fields of biology continued, and these advances in turn have been assimilated into the palaeontological perspective, but the mode of operation of the science was well established by the close of the 19th century. Detailed investigations into the process of evolution as reflected in the rocks are continuing, and many of the questions on the evolutionary origins of living animal and plant groups are as hotly debated today as they ever were. But few people doubt the overriding principle of evolution in the shaping of the biological world.

TURNED TO STONE

When fossils were regarded as curious freaks of nature, explanations of the way they were preserved could be fanciful. Ammonites found at Robin Hood's Bay on the Yorkshire coast, UK, were supposed to have been snakes turned to stone by St Hilda. Enterprising local craftsmen embellished the fossils by carving on an appropriate snake's head. As it became clear that fossils were present in the rocks as a result of natural processes, the mechanism of fossilisation became better understood. The majority of fossils comprise the hard parts of the animal or plant, or structures resistant to decay. Shells, bones, wood and teeth are all likely to be preserved as fossils. Molluscs and mammals are thus likely to have a good fossil record,

BELOW: The ammonite *Dactylioceras commune*, from Jurassic rocks, Yorkshire, England, with a snake's head added by a sculptor.

ABOVE AND BELOW: The Silurian graptolite *Dictyonema retiforme*, 425 million years old. *Dictyonema* formed a cone-shaped colony in life, and is flattened from the side (top, from an area between Niagara Falls and Lewiston, New York, USA) and is flat and from above (bottom, Hamilton, Ontario, Canada).

worms and amoebae a poor one. The clusters of dead shells one finds in rock pools on the beach – limpets, crab claws, a winkle or two, a broken sea urchin – are typical of the sort of debris that may become fossilised. Many fossils result from the cast off, outgrown or broken shells of marine animals. Such shell material becomes covered with sediment and can be said to be a fossil from that stage on. Many fossils more than 1 million years old can look remarkably like shells picked up from the beach today. Most shells are porous and frequently the small pores in the fossilised material become the site for deposition of minerals, which make the fossil more dense than the original shell of the animal.

As sediments pile up at the site of deposition their coherence increases, some of the water they contain is expelled, and many subtle chemical changes occur to produce rock types such as shale, sandstone and limestone. In the harder rocks, such as limestone, fossils tend to retain much of their original shape and convexity, but shales frequently become compressed, and as this happens the fossils become flattened. Sometimes the same fossil species can have a very different appearance according to whether it has been preserved in hard rock or flattened. Further changes can happen to fossils as a result of their sojourn in the rocks. Most shells are made of calcium carbonate, which is soluble in carbonated water. In porous rocks, especially sandstones through which groundwater travels, the shell can be dissolved away. The fossil is not lost in this case because the rock itself will have taken an impression of the shell, just as a fingerprint can be impressed into clay. When such a rock is split open the inside of the shell will be preserved as an internal mould, while the other half (the counterpart) of the specimen will contain the impression of the exterior details. To obtain a reconstruction of the whole shell you need both halves of the fossil specimen. The golden rule of collecting is: never throw away the counterpart. Many a vital specimen has gone tumbling down a scree slope when this rule has been ignored. In some cases the cavity left after the shell has been dissolved is replaced by another mineral; if you are very lucky the mineral might be opal.

Sometimes these secondary replacements can be advantageous to the palaeontologist. Fossils with shells originally of calcium carbonate may be replaced by the mineral silica. If this happens in a limestone, the enclosing rock can be dissolved in acid and, because the silica is insoluble, the replaced fossil will remain. In this way remarkably delicate details of the original fossils can be preserved, such as spines and other features, which are impossible to dig out

BELOW: Internal (left) and external (right) moulds of the Ordovician trilobite *Placoparia* from the Czech Republic.

mechanically. A rarer type of preservation is perfect replication almost molecule by molecule of the original fossil, so that minute details of the microstructure are determinable. Such petrifaction is usually in fine-grained silica. One famous example is the Rhynie Chert, Scotland, a Devonian petrifaction of some of the earliest land plants. The perfection of preservation is such that sections of these early plants clearly show the individual cells in the plant tissues. Such specimens are of immense importance in revealing the most intimate details of the structure of extinct organisms, which can then be compared with some confidence with their living relatives for which there is complete information.

RIGHT: Silicified brachiopods showing the inside of the dorsal valve in the upper picture, and the outside of the ventral valve in the lower, with long, delicate spines which would normally break off in the rock.

GEOLOGICAL MIRACLES

In very rare cases it is not only the hard parts of animals and plants that are preserved in the fossil record. Such geological 'miracles' also preserve the impressions of the soft organs of animals that would normally decay without trace. There are always special geological circumstances in such cases. One of the most famous, and oldest, of these 'miracles' is the Burgess Shale from British Columbia, Canada. This is a black, fine-grained Cambrian shale about 515 million years old, with a host of wonderfully preserved fossils, many of which are unknown anywhere else in the world. It affords us a unique glimpse of almost the whole spectrum of life at this very early stage in its history, and shows many animals not preserved under normal circumstances in the fossil record. Trilobites and other arthropods are preserved with all their limbs, antennae, and even their gut contents. Some of the animals are hard to match with any living organisms, and may represent kinds of creatures that have long since vanished from the Earth. Many types of worms are present in the fauna of the Burgess Shale. The soft parts have been preserved as the thinnest of films, and the worms were probably buried rapidly, before the soft parts could decay, in an environment where they could not be shredded to pieces by scavenging organisms. A similar occurrence in the Devonian Hunsrück Shale, Germany, has the soft parts of animals preserved as a thin film of the mineral iron pyrites. In this case the structure of the soft parts can be studied using X-rays, which pick out the iron pyrites; thus the structures can be photographed in the rock even though they are not visible on the surface.

The early bird *Archaeopteryx* is preserved in a fine-grained limestone, creamy brown in

ABOVE: *Marrella*, from the Cambrian Burgess Shale, a peculiar primitive arthropod unknown elsewhere.

LEFT: A phacopid trilobite from the Devonian Hunsrück Shale, Germany. Fine details of the limbs are preserved.

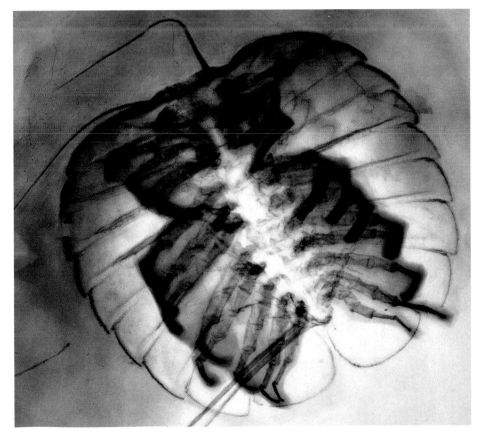

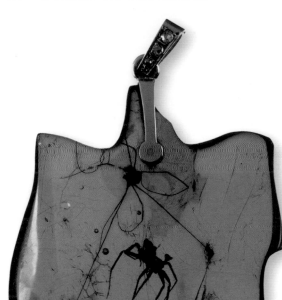

ABOVE: Insects and spiders, not normally preserved as fossils, can be trapped in amber as old as 120 million years. Top left: a pendant containing two arachnids, a harvestman *Phalaphium* sp. and a lynx spider *Oxyopes* sp., top right: a pendant with a spider and a cricket; bottom: ring with a long-legged fly from the family Dolichopodidae dating from about 35 million years ago.

colour, and used for the manufacture of lithographic blocks, found near the Bavarian town of Solnhofen. This could scarcely be more different in appearance from the Burgess Shale, but like it the lithographic limestone retains spectacular remains of a whole host of animals with scarcely any fossil record elsewhere. The fossils are Jurassic in age, and many are related, distantly, to animals still living. Besides *Archaeopteryx* there are reptiles, dragonflies, relatives of the horseshoe crab *Limulus*, sea spiders and mammals, in effect a mixture of terrestrial and marine life. The Solnhofen deposits are believed to have built up on the boundary between land and sea, probably in a lagoon, where a sticky, limy mud was accumulating. Flats of this mud were probably exposed at low tide, and at this time *Archaeopteryx* became entrapped. The fine mud was ideally suited to entomb the remains of the bird and also to take an impression from delicate feathers that would otherwise have decayed without trace. It is fortunate that this happened, because otherwise there would be doubt as to whether or not *Archaeopteryx* was really a bird. Some scientists have claimed in the past that the *Archaeopteryx* specimens are fakes, manufactured by impressing feathers of living birds on a prepared surface surrounding genuine dinosaur fossils. This claim is inherently improbable given the discovery of fossils of *Archaeopteryx* from time to time over many decades (it would imply a scientific conspiracy of massive proportions, and without commensurate motivation). Furthermore, careful examination of the fossils has not revealed any evidence of such fakery. Indeed, new studies using CT scans of the inside of the skull have even been able to reconstruct the brain case of this important animal, showing that it had a bird-like anatomy.

Rare, geological miracles like the Burgess Shale and the Solnhofen limestones are of exceptional palaeontological importance; the information they yield is like a floodlight on the past, whereas most geological sites are more like an intermittent flashlight. The geologically youngest of such exceptional fossils are the frozen mammoths of Siberia, dating from late in the last ice age. These extinct giants were apparently frozen so quickly that their hair and flesh are preserved almost as if they had been prepared for the supermarket. Speculations about regenerating one of these remarkable animals from their frozen cells are probably over-optimistic.

Amber ornaments were extremely popular in Victorian times, and the most prized of these had small insects displayed within the amber droplets. Apart from its beauty, amber preservation is another exceptional occurrence where animals not normally preserved as fossils are found in abundance. Amber is generally 70 million years old or less (although some is as old as 120 million years). The majority of the insects encased within it are related to living forms and are of great importance in understanding the genesis of the most diverse group of living invertebrates. Amber is produced as a resin oozing from the branches and trunks of coniferous trees. Insects and spiders became trapped in this resin, and enclosed within it as more resin was added and solidified. Hardened resin is extremely tough, and so has a high chance of being fossilised, carrying with it its cargo of preserved insects. Lumps of amber eventually found their way into sediments, from which they can be recovered like any other fossil. The same process of entrapment still goes on today, and the Recent (geologically speaking) hardened resin known as copal has been sold as ersatz amber. You cannot crack out amber insects from their casing – they simply fall into dust. However, since 2000 a new application of scanning at the European Synchrotron Radiation Facility in Grenoble, France, has managed to capture perfect images of entombed fossils inside amber, even if the amber is opaque to light. These include tiny ants, which can be examined in minute detail. The same technique has been applied to study very ancient fossil embryos.

TRACE FOSSILS

A fascinating branch of palaeontology is concerned with the traces left by the activity of extinct animals – trace fossils (ichnofossils). These can be footprints, like the tracks of dinosaurs, left by animals as they crossed the sediment looking for food, or burrows made by animals escaping from a predator or laying their eggs. In many cases the animal itself is not preserved, and although it may be possible to deduce what it was doing, it is often not possible to say what animal was responsible for making the track. Huge, three-toed tracks of dinosaurs can tell us a lot about the gait of the animal that made them, or for example what its stride was, or whether the front limbs touched the ground. However, many tracks are inconspicuous, such as labyrinthine or braided paths made by worms, and some of these worm 'diggings' are the only fossil record we have of these creatures.

Trace fossils are abundant even in the Early Cambrian. They are especially numerous in sandy rocks, which otherwise lack body fossils. One of the Cambrian occurrences is the famous 'Pipe Rock' of the northwest Highlands, UK: the pipes that give the rock its name are closely packed straight tubes which were made by some sort of worm. Other beds of rock in the same formation contain U-shaped or funnel-shaped tubes. The tracks made by trilobites are numerous in rocks of Cambrian and Ordovician age. These include winding trails and short burrows, some of which contain the impressions left by the legs of the trilobites. Trace fossils are given names, just like body fossils; the trilobite tracks are called *Rusophycus* or *Cruziana*.

An explanation is needed about the preservation of tracks: most tracks are dug into a sediment surface, and an overlying layer of sediment then fills up the tracks, so that the cast formed by this infilling is a positive impression of the track itself. Burrows are dug deep into the sediment, and they often fill, once vacated, with more sediment of a different colour. Collecting past tracks can be almost as instructive as collecting the fossils of shells. Different kinds of animals live in different environments, and leave differing evidence of their activities. By studying tracks it is possible to find out a lot about the animals that lived on the former sea floor, even without remains of the animals themselves. Tracks should be distinguished from borings. Borings are made into a hard medium, such as rocky surfaces or wood. Some bivalved molluscs specialise in boring into hard rocky surfaces, and these too can be found as fossils, usually sitting at the end of their self-made caves.

OPPOSITE: This Middle Jurassic 147 m (482 ft) long sauropod trackway (175–165 million years old), from a site in Portugal, is the longest trackway on record. The track was made by an animal walking in a straight line across a muddy estuary.

BELOW LEFT: Upper Cambrian trilobite tracks named as *Cruziana semiplicata*, thought to be made by a trilobite called *Maladioidella*, from Oman.

BELOW RIGHT: A broken limestone pebble revealing fossils of marine bivalve molluscs, *Pholas* sp., in their flask-shaped borings. This specimen is from the Late Eocene, Hampshire, England.

HISTORY OF FOSSILS WITHIN THE ROCK

Once they are incarcerated in the rocks, fossils are passive passengers, and what subsequently happens to the rocks also affects the fossils themselves. Not all rocks lie undisturbed until the hammer cracks open the booty they contain. Initially the fossils usually lie parallel to the bedding planes – the surfaces parallel to the surface of deposition. Some fossils remain in this attitude as the rocks are uplifted above the former ocean bed, to become exposed in cliffs, quarries or cuttings. More usually the uplift process, which may be connected with Earth movements (see Chapter 2), also produces tipping and gentle folding of the rocks, so that the bedding planes are no longer horizontal. This does not affect the fossils, which can be collected in the usual way. Where the Earth movements are more violent, the rocks may be squeezed and distorted, and so are the fossils contained inside. This is particularly the case in soft rocks, like shales. Older fossils are more likely to be found in this condition, because they have had more time to be involved in violent events, but many have escaped such action, and are as well preserved now as they were when they were first fossilised. Distortion includes stretching and twisting, and this results in a lot of the finer details of fossils being destroyed. Palaeontologists would like to have perfectly preserved material to work from, but often distorted fragments are all that is available from a large area, and they have to be identified as best they can.

BELOW: Folded rocks resulting from tectonic processes may contain fossils that have been distorted from their true shape.

The process of distortion does not stop with stretching and bending. As the rocks are squeezed progressively they sometimes become heated strongly and under these conditions the rock itself begins to change. This can have the effect of removing the fossils completely, the small ones first, then the most robust. Even so, fossils have been known to survive the most intense heating and high pressure. A very common secondary change that occurs in shaly rocks under pressure is that they develop a cleavage. This means that they start to split, not along the bedding planes on which the fossils lie, but at a high angle to the bedding, sometimes at right-angles. The kind of black or purple slates used for roofing are cleaved rocks of this kind. You can often see stripes passing across such slates and these are sections through the original bedding. It is no use splitting open such rocks in the hope of finding fossils parallel to the cleavage, although you might see a section through one if you are lucky. Fossils can be recovered with much hard work by smashing the slates in such a way that part of the original bedding is exposed. They are usually in a rather sorry state by the time they are found.

BITS AND PIECES

Many fossils are only fragments of the whole animal or plant. To piece together the whole organism is rather like doing a jigsaw puzzle without the benefit of the complete picture to work towards. Piece has to be added to piece, and the larger and more fragmentary the animal the more the result will be in question. Not surprisingly mistakes have been made. The first reconstruction of the dinosaur *Iguanodon* finished up with its thumb on its nose! Trilobites are much more commonly found as pieces rather than as whole animals, and many kinds are known only from heads or tails until one day a lucky collector turns up a whole one. The problem is particularly acute for large plants, because there is nothing very obvious to connect the root with the trunk, or the trunk with the leaves if they are preserved in different places (and they usually are). Flowering structures, not being easily fossilisable, are sometimes even more difficult to assign to the plant. The result is that different names are given to different organs, one for the root, one for the bark and so on. Eventually, when the links are made, one name suffices for the whole plant.

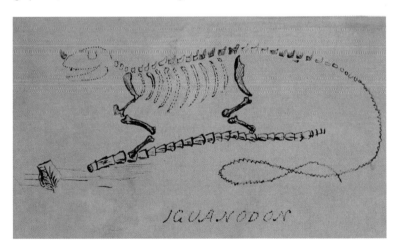

There are many problems with piecing together fossils: distorted fossils have to be restored to their original shape, left valves of clams have to be matched with right, vertebrae and limb bones need to be placed together in the correct order. Many fossils are only known from fragmentary material, and it may be years (if ever) before the next piece is discovered. Just as there are rare and abundant species today, so there are rare and common species of fossils. Some species are so abundant that one is almost certain to turn up a specimen if rocks of the right age are hammered; others are so rare that a collector may go to a quarry one day and find a specimen but visit the same place for years afterwards without finding a second one. This is all part of the particular fascination of palaeontology; you never know

ABOVE: Gideon Mantell's original restoration of *Iguanodon*. Based on the skeletal anatomy of a modern lizard. Mantell placed the spike-like thumb of *Iguanodon* on its nose, like the horn of a rhinoceros. This misconception persisted for many years.

quite what will turn up. The discovery of new kinds of fossils is a regular occurrence, even in rocks that have been investigated by many collectors before. Each new discovery patches in a little of the story of life that was previously obscure. Even now many parts of the world are being explored for the first time, and the new palaeontological finds show little sign of abating. For example, at the end of the 20th century discovery of 'feathered dinosaurs' from China quite altered our picture of what we previously thought were well-known fossils; it all depended on new and better preservation. Professional palaeontologists often feel swamped by the sheer variety of different fossils there are to deal with.

OPPOSITE: Excavation of the dinosaur *Baryonyx walkeri*, at Ockley brick pit, Surrey, England in 1983.

Although they are fascinating just as attractive objects, fossils are of enormous practical importance in interpreting the history of the planet as a whole. The study of fossils links in with other branches of geology and biology; indeed a knowledge of related sciences is essential to appreciate the significance of fossil finds. Recently, for example, the study of 'chemical fossils' has become an important new field. Such chemical fossils are often the remains of decayed organic material, which provide enduring organic molecules left in the rock, and these can tell us what kind of organism made them. Study of the variations in the isotopes of carbon and oxygen can elucidate past atmospheric and oceanic conditions. Sophisticated equipment capable of measuring minute quantities of organic material is becoming more and more important to the professional palaeontologist. The next two chapters will explore some of the connections with other geological sciences, to show how the animal and plant life of the Earth is intimately bound up with the story of the rocks themselves, and the configuration of the continents and oceans. As Lyell recognised, the forces of physics have been the same through geological time, so the processes that formed the present Earth are not beyond our understanding. But the unchanging physical laws operate on a thoroughly mutable world, and the arrangement of land and sea has changed repeatedly. For hundreds of millions of years living organisms have altered in harmony with the world, and in the process have themselves transformed it.

ABOVE: Sir Charles Lyell (1797–1875), who believed that the Earth had been subject to gradual change through the action of processes such as earthquakes and erosion.

Time and change

The Earth is 4,550 million years old. This figure is now generally accepted, but has only become so in the last few decades, especially since the age of the moon was determined. If we return to the 19th century, at the time when Hutton saw that the Earth must be 'immeasurably' old for present processes to account for all the varied features of the rocks and their immense thickness, the business of putting an actual age on the Earth was very problematic.

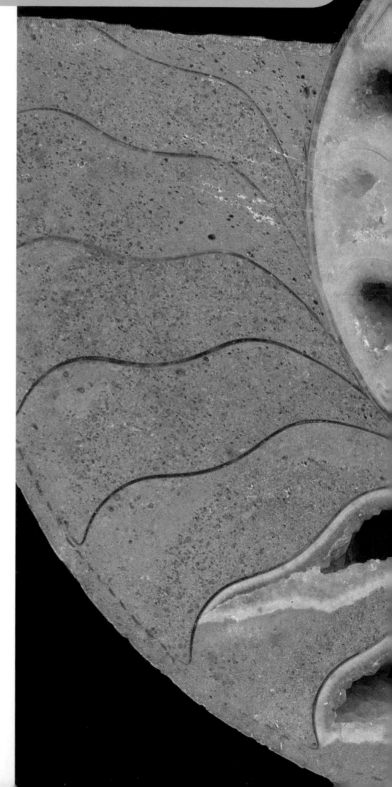

RIGHT: An ammonite, *Brasilia bradfordensis*, which has been been sectioned to show its chambers with calcite crystals formed in some and hardened mud in others.

Once the idea of Creation within a few thousand years disappeared, it was obvious that the Earth had to be much older. But how much? Millions of years certainly, but 10 million, 100 million, or 1,000 million? The fact is that it is impossible for the human mind to grasp such lengths of time as these. It was easy enough to realise that a long span of time was needed, but difficult to devise methods of assessing the actual amount of time involved.

GEOLOGICAL TIME

The history of the Earth is written in the rocks, and the rocks can be divided into natural units. Once the sequence of rocks and fossils contained within them was worked out, it was possible to say something about the different kinds of organisms that had populated the Earth and replaced one another over geological time. In the 19th century, fossils were found in all rock units except the underlying Precambrian rocks, which seemed to be barren of all fossil life. How much of the Earth's history did these represent? At least one school of thought believed that these ancient rocks were partly the original crust of the Earth (many of them being crystalline) and an approximation of the age of the Earth could be obtained by guessing the length of time taken for the deposition of the Cambrian and later rocks. If the rocks accumulated at a fixed rate, then the maximum total thickness of the rocks should provide a basis upon which to measure the time it had taken to deposit them. At the time the subject was debated in various ways, in particular focusing on the suggestion that the rocks would be compressed by burial, and the resulting estimates of total deposition time ranged from less than 20 million years to more than 700 million years. The great physicist, Lord Kelvin (1824-1907), assumed that the Earth had cooled from a totally liquid state, and the time taken to produce the current condition of heat flow and solid crust would be between 20 and 40 million years. By the 1890s this seemed to be the most scientific answer.

At the same time the knowledge of the relative timescale of geology had come to resemble closely the scale we use today. The question of absolute (in millions of years before present) age is a different one, and the whole edifice of geology was built largely on relative relationships. The Silurian rocks were older than the Devonian rocks, because they underlay them, and the fossil fauna could be shown to be different; the Devonian in turn underlay the coal-bearing Carboniferous, and these rocks underlay the Permian formations, and so on up the geological column.

The fossil faunas also seemed to change in a radical way from one of these broad units to another, and what better way, therefore, to define the divisions of geological time? The development of the rock formations varied from one area to another, and in some cases provided the name of the geological period; for example, the Jurassic period was named after the Jura mountains. Some of the natural divisions of geological time were bounded by unconformities, angular breaks in the sequence of rocks. Rocks lying beneath an unconformity were uplifted and folded before the rocks lying on top of them were deposited. The divisions of geological time from the Cambrian onwards were established in Europe, and it is a measure of the acumen of the geologists who established them that the original concepts still survive today.

The major divisions of Phanerozoic geological time (periods or systems) are shown on p. 32. The first four of these (Cambrian, Ordovician, Silurian, Devonian) all take their names from British localities; Cambrian from the Latin word for Wales, Ordovician and Silurian from

two of the old Welsh tribes, and Devonian from Devon. The Carboniferous rocks comprise the coal-bearing strata, and the Permian rocks are typically developed in the Perm mountains of Russia. Triassic rocks naturally fall into three (tri-) divisions in their characteristic development in Germany, and the Jurassic period was christened as mentioned above. The Cretaceous included the pure white limestone known as chalk (Latin: *creta*). For broader purposes it is still useful to talk about eras, three great divisions of fossil-bearing time: Palaeozoic (literally 'old life' encompassing the Cambrian to Permian periods), Mesozoic ('middle life' spanning the Triassic to Cretaceous) and Cenozoic (or Tertiary) taking us up to the base of the Pleistocene.

The boundaries between the eras represent the most important extinction events the world has known, when much of the living world was replaced by new and different organisms, a graphic way of dividing the history of life into three portions. We can now add a fourth era, the Proterozoic (ancestral life) for the later part of the Precambrian, from which fossil organisms are widely known, the topmost part of which, the Ediacaran, is the most recent addition to the catalogue of names for geological time. The Cenozoic periods are shorter than the preceding ones, and were originally proposed by Lyell on the basis of how closely the fossil faunas resembled those of the present day. They include the Eocene ('dawn of the present'), the Miocene ('less than present') and so on. Finally the Pleistocene ('nearly present') includes the ice-formed deposits that have been draped like an irregular blanket over the earlier geology of the northern hemisphere, where the geological record merges into the history of humans.

Once the language of geological time had been developed it was a great help in communicating the discoveries that had been made internationally, and in building up the first pictures of whole faunas for particular periods. Certain periods acquired popular descriptions, some of which seem to have stuck. The Devonian became the 'Age of Fishes', the Carboniferous the 'Age of Amphibians' and the Jurassic– Cretaceous the 'Age of Reptiles'.

ABOVE: An unconformity is a break between two rock types laid down at different times. The rock here shows an angular unconformity, in which rock layers are at different angles; Kilkenny Bay, north Somerset, England.

RIGHT: The main divisions of geological time (figures are in million years). The geological timescale was originally established from the relative ages of the rocks, partly based on fossils. Some of the important fossils for dating are shown.

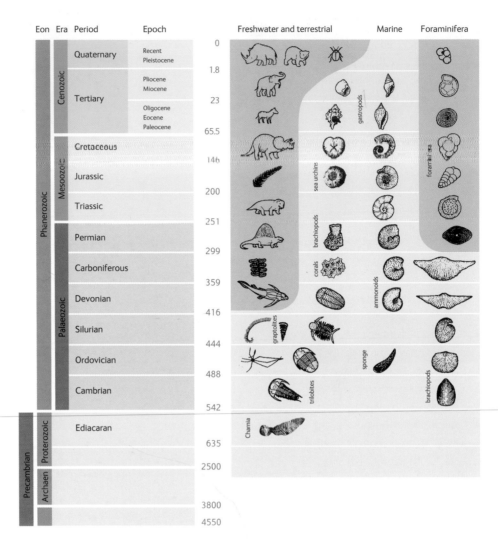

Eon	Era	Period	Epoch		Freshwater and terrestrial	Marine	Foraminifera
Phanerozoic	Cenozoic	Quaternary	Recent / Pleistocene	0			
				1.8			
		Tertiary	Pliocene / Miocene			gastropods	
				23			
			Oligocene / Eocene / Paleocene				foraminifera
				65.5			
	Mesozoic	Cretaceous		146		sea urchins	
		Jurassic		200			
		Triassic		251			
	Palaeozoic	Permian		299		brachiopods	
		Carboniferous		359		corals	ammonoids
		Devonian		416			
		Silurian		444	graptolites		
		Ordovician		488		sponge	brachiopods
		Cambrian		542		trilobites	
Precambrian	Proterozoic	Ediacaran		635	Charnia		
				2500			
	Archaen			3800			
				4550			

These tags have endured, and they do serve a purpose to emphasise some of the largest animals that lived in the respective periods, but they also give a rather false impression of all the different biological events that were going on at the time. There is always a temptation to view the fossil record as if it were a kind of staircase with progressive steps leading upwards towards humans. This is misleading, because evolutionary activity has been unremitting in even the humblest of creatures, and dominance of the natural world by larger animals is only a matter of their conspicuousness.

The geological periods were soon to prove only the most general way of subdividing geological time. The periods themselves could be subdivided into segments, which would enable a much more refined way of talking about the relative ages of organisms. For example, dinosaurs changed greatly during the Cretaceous period, and it was necessary to have a way of describing the timing of these changes. Periods were divided into early and late parts (mid-parts in many cases as well). The finest subdivision became (and remains) that of the zone. A zone (or biozone) is a division of geological time characterised by a particular assemblage of fossil species. It is a small segment of geological time through which the evolutionary histories of various organisms pass. There will be a number of animals or plants that are unique to a specific zone, although others that have evolved more slowly may range through more than one zone. The name of the zone is taken from one of the most characteristic of its defining organisms. For example, in the Ordovician, graptolites are of importance in subdividing the rocks, and the 'Zone of *Nemagraptus gracilis*' is named after one species that is of widespread occurrence in its zone, although it is accompanied by other species characteristic of the same time period. The zone is another way of communicating the exact age (on the relative timescale) of a fossil, and, one hopes, means the same in all areas of the world. Some kinds of organisms have become more useful in the definition of zones than others, and the most useful have proved to be those which evolved rapidly (because they enable the time to be sliced finely) and which achieved wide geographic dispersal.

ABOVE: The graptolite, Ordovician *Nemagraptus gracilis*, is abundant in rock of a particular age. It is known as a zonal fossil: it identifies a particular age of rock throughout the world.

Much of the attention of geologists is devoted to trying to establish the age relationships between strata from widely different localities – the correlation of rocks – and this is the cornerstone of the branch of geology known as stratigraphy. It is not necessary to know the exact age in millions of years to be able to correlate; the basic statement is "this rock is the same (or not the same) age as that one" whether the age of the rock in question is 2 million or 200 million years old. The fossil content acts as the clock.

The kinds of fossil organisms used as the basis for zones vary through geological time, as one group rises to prominence only to be replaced by another. It is the humblest of fossils that are often the most useful as the basis of the zonal schemes. In theory any organism with a good fossil record can be used for a zonal indicator, but obviously dinosaurs are far too large to be recovered from an average roadside exposure even though they may have evolved very quickly. Common fossils like ammonites, brachiopods and trilobites are used as zonal fossils, partly because they can be recovered from most (marine) sediments of the right age, and partly because they show enough variation through time to be readily recognised in the laboratory by the palaeontologist.

A whole separate branch of palaeontology has grown up around using the smallest of fossils as zonal indicators. This is known as micropalaeontology and is described further in Chapter 9. Small fossils are of particular use in dating rocks recovered from boreholes, where the narrowest of cores may yield large numbers of diagnostic fossils. Not surprisingly, this kind of palaeontology is much employed by oil companies and other commercial enterprises concerned with recovering mineral wealth from considerable depths. Dating the rocks, and correlating between boreholes, is a fundamental part of mineral and oil exploration. It is possible for several zonal schemes to exist side by side, referring to the same time period; one may be primarily concerned with microfossils, another with brachiopods or ammonites and so on. One system usually becomes the 'standard' for a particular geological period, often the one first proposed.

The most important fossils used as the basis of zones change through the geological column. In the Cambrian, trilobites are more widely used than any other group, largely because they were among the most varied and numerous of the invertebrates at the time. For the Ordovician and Silurian periods, graptolites have been widely employed for zonal purposes, although trilobites, brachiopods and other fossils are still used in rocks where graptolites are absent. Before the end of the Devonian the ammonites had evolved, and they are probably the single most important zonal group from the Carboniferous through to their extinction at the end of the Cretaceous, although their use is often supplemented by other invertebrate organisms. Micropalaeontological zonations become progressively more important through the later part of the Palaeozoic and the Mesozoic, and the use of small, unicellular microfossils known as foraminifera outweighs that of other kinds of fossils in the Tertiary. Zones are capable of dividing time into fine slivers.

It is estimated that at best a zone (or its ultimate subdivision, a subzone) can divide time into slices of about half a million years or even less, a remarkably precise calibration.

ABSOLUTE AGES

BELOW: Time measured by radioactive decay. The half-life is when half the original radioactive element has decayed away.

Obtaining an age for rocks in millions of years is now a routine geological method, and provides more accurate data that help answer the question of how long the processes shaping the Earth have taken. The method of determining such an age depends on the varieties of elements known as radioactive isotopes. These are unstable forms of elements that decay (change) slowly into other elements, in the process giving off radioactive emanations like gamma rays. The rate at which this happens is controlled by a simple natural law, so by knowing the amount of the parent element left, and the daughter element generated, it is possible to calculate the time taken to produce what we see today. The radioactive clock, unlike fossils, converts immediately to millions of years. The method depends on accurate measurement of numbers of atoms, and techniques for doing this have improved markedly over the years, thus making the results ever more reliable. The radioactive clock is often set to zero when a liquid rock cools below a certain point. Many radiometric ages are obtained from igneous rocks, which of course do not contain fossils, so the method often operates where fossils cannot be of direct use.

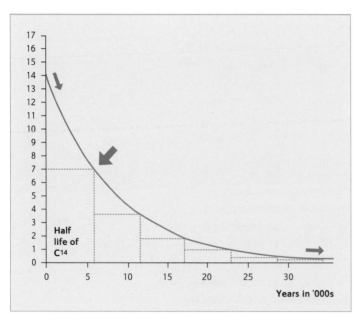

The methods originally employed depended on a few of these natural transformations, particularly entailing one isotope of uranium changing to another of lead. More and more methods are in use today, and these can now be used on sedimentary as well as igneous rocks. They are often referred to by the names of the mother and daughter elements, such as rubidium– strontium (Rb–Sr) dating, or potassium–argon (K–Ar) dating. The rates at which the decay occurs vary from one natural reaction to another; the slower the process the more useful that reaction is for dating extremely old rocks. Radiocarbon dating, involving the change from carbon-14 to nitrogen-14 happens relatively quickly (geologically speaking), and this method is particularly useful for dating young events measured in thousands of years. This method, particularly with modern refinements, can give a resolution that exceeds that of any other technique, and is often the only recourse for dating isolated sites (like the abandoned camp fires of our forebears). At the other end of the scale the application of radiometric dating techniques has revolutionised our understanding of the huge areas of Precambrian terrain from which fossils are unavailable. Crystals of the stable mineral zircon provide particularly useful dates. These survive as 'time capsules' from the early days of the Earth and have identified the oldest rocks on Earth as about 4 billion years old in Greenland and Western Australia. The dating of meteorite and moon rock has provided us with the age of formation of the Earth itself.

Fossils are best utilised for relative dating in the time period between the Precambrian and Pleistocene, and this is the timescale with which this book is primarily concerned. Here ages derived from radioactive decay of elements provide a scale into which the stratigraphic ages of fossils can be linked. Radiometric ages always carry a margin of error ($\pm$) of some magnitude, though this is always diminishing as methods become more accurate. The relative timescale provided from fossil evidence is still useful. The two methods are complementary; fossils are used in the fine division of periods, absolute ages provide a guide to how long ago events took place.

The cooling ages in igneous rocks give the time at which particular bodies, for example of granite, were intruded into the surrounding rock. Where rocks have been reheated and folded during mountain building events, this too can 'set' the radioactive clock, providing a method of determining when major phases of metamorphism occurred. All these dates can then be placed into the geological timescale, which provides the general framework for the timing of events. The combination of absolute ages and the relative stratigraphic scale is one of the major achievements of geology.

MAJOR CYCLES IN EARTH HISTORY: MOVING CONTINENTS

The evolution of life cannot be separated from the evolution of the planet, which is its cradle and its grave. It is now certain that the surface of the Earth itself has changed its configuration several times. The Earth's lithosphere is a thin skin, of almost negligible thickness compared to the mantle that lies beneath it. The tectonic plates forming the surface of the Earth move, and, as they do so, continental geography changes. This happens at an immensely slow rate (1–2 centimetres or about ½–1 inch, per year) resulting in what was once called 'continental drift', but cumulatively the effects are profound. For example, at present Africa and South America are moving further apart. The blocks of continental crust that form these huge areas are thick, tough and relatively stable. The crust that floors the oceans is relatively thin and mostly composed of basaltic volcanic rock. As the continents draw apart new oceanic crust in the form of volcanic rock wells up from the mantle along the mid-ocean ridges. The Earth's crust is composed of a number of more or less rigid plates, moving past, towards or away from one another. As new crustal material is generated along the mid-ocean ridges, so other parts of tectonic plates must be destroyed elsewhere. Thin oceanic crust is removed along subduction zones, where it plunges beneath the relatively rigid continental blocks. An important subduction zone lies off Japan today. This area is one of sudden earthquakes, for the downward slide of oceanic crust is far from smooth. The physical expression of these downward-diving plates is found in deep ocean trenches offshore. Along the line of origin of new oceanic material, volcanoes may achieve sufficient height to break above the sea surface as volcanic islands. A different kind of volcano may originate along the line of plate consumption, as the plates plunge down and melt at depth.

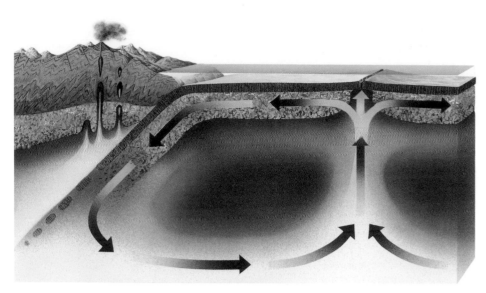

RIGHT: Generation of new oceanic crust occurs at the mid-ocean ridges, and destruction of the oceanic crust takes place at the continental margin.

The geophysical basis of plate movements is now being fully explored, although, oddly enough, it was the supposed lack of an appropriate physical mechanism that made the scientific establishment sceptical of the whole idea of continental drift in the early 20th century. As far as the palaeontological world is concerned, the record of fossils was one of the first compelling bodies of evidence to show that continental drift was a reality. For example, it was observed that the Late Palaeozoic fossil plant floras with *Glossopteris* were remarkably similar in India, southern Africa and South America. Subsequently the same peculiar plants turned up in Antarctica. How could one explain such a distribution with the continents as they are at present? It seemed unlikely that these terrestrial organisms could have drifted across such wide ocean barriers, and the idea of thin land bridges connecting widely separated areas seemed preposterous. However, if one fits the continents back together to form Gondwana the *Glossopteris* floras are brought into relatively close proximity, and none of these objections apply. The flora looks like a relatively continuous belt occupying what were cooler latitudes at the time. The same arguments were applied to terrestrial vertebrate fossils. Of course animals and plants are capable of crossing bodies

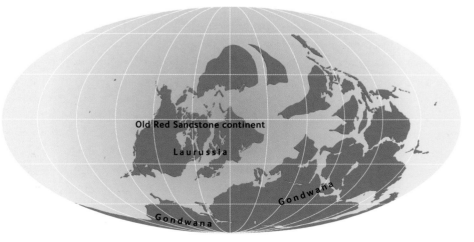

RIGHT: Reconstruction of the continent Gondwana about 409 million years ago, which broke up to give the present day continents.

of water, but a look at the animals that have managed to reach remote islands today demonstrates the fact that only a few kinds of animals succeed in doing so. The oceans are effective barriers.

The combination of faunal, floral, geological and geophysical information led inevitably to the conclusion that the present continents were fused together in one great 'supercontinent' in the Permian. This is called Pangaea. In the Mesozoic Pangaea broke up into pieces that moved apart, eventually to give the present continents. These separated slowly, with widening oceans between them, a process that continues to the present day. This means that the basaltic sea floor of the Atlantic and Indian Oceans must have been created at the ocean ridges since the Jurassic. To prove this, no fossils older than Jurassic are known from the sediments which have accumulated in these regions, or indeed in any region of the sea floor. All sea floors are, geologically speaking, young rocks, and the youngest of all are still being created in volcanic eruptions like the one that built Surtsey (off Iceland) in a matter of days.

In some cases, the drifting continental blocks, or the island arcs flanking them, collide with one another. The most impressive results of this are mountain chains like the Himalayas, which were thrown up by the collision of the Indian subcontinent with the main continental landmass of Asia. Such effects are highly dramatic, but the elevation of the Himalayas (which continues) has taken millions of years, and has involved several phases. The squeezing of the rocks along linear mountain chains results in their being folded, dislocated and maltreated in many other ways which are a delight to a geologist, but sometimes a cause of gloom for the palaeontologist, because the fossils in the rocks are subjected to the same treatment, and may emerge at the other end much the worse for wear. The whole process of squeezing culminates in heating, metamorphism and sometimes melting of the rocks, which effectively destroys the fossil record, although there are some remarkable examples of fossils having survived enormous temperature elevation.

As Pangaea broke up, the dispersing continents carried the fossil remains of the Permian–Triassic supercontinent to their present scattered positions. The animals that lived on the drifting continents thus became isolated from their contemporaries in the rest of the world. For terrestrial vertebrates such isolation can result in the marooned animals having their own independent evolutionary history. Australia separated from Pangaea and carried eastwards a cargo of animals that eventually became isolated from further contact with the more advanced mammals that came to dominate the rest of the world. The early migration routes into Australia were through South America and Antarctica. The eventual isolation did not stop evolution, quite the reverse, and the marsupials evolved into a very varied group of animals. They were able to occupy almost all the ecological niches that were available to them, from burrowing, tree climbing or grazing to carnivorous or scavenging

BELOW: When landmasses collide, large areas of the sea floor are folded and uplifted. Marine fossils, like this ammonite *Virgatosphinctes* found at an altitude of over 5000 m (16,405 ft) in the Himalayas, may be raised to great heights far from their origin.

RIGHT: As Gondwana and
Pangaea broke up and the
resulting continents moved apart,
animals living on the continents
such as South America became
isolated, and another set of
endemic mammals evolved.

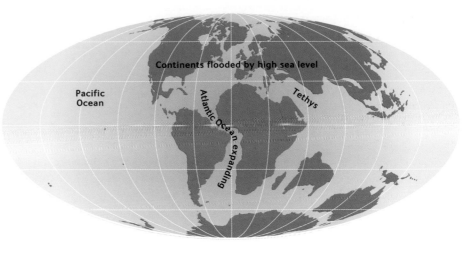

Continents flooded by high sea level

Pacific
Ocean

Atlantic Ocean expanding

Tethys

habitats. The primitive pouch is a common feature linking all of these animals, from the marsupial mouse to the great grey kangaroo. Some extinct giant marsupials died out after humans first settled in this self-sufficient world. The subsequent introduction of the domestic cat has done a lot of damage to smaller marsupials; it does not take long to undo what tens of millions of years of plate tectonics have created. South America was similarly isolated until the recent geological past and another set of endemic mammals evolved, including the giant sloth (*Megatherium*) the bones of which Darwin collected in his voyage on the *Beagle*.

Plate tectonics has also had its effects on marine organisms. The free-floating larvae of most marine organisms mean that oceans are not the barriers for sea creatures that they are for marsupials. Marine animals are, however, adapted to particular water temperatures, which is why the species of molluscs in the tropics are generally different from those found in shelf seas that surround the North Atlantic. The distribution of marine fossils can be used as a kind of thermometer to show how the water temperatures have changed as the continents, with their fringing seas, have changed position relative to the lines of latitude. In some cases the continents themselves act as a barrier for the marine animals. The general north–south direction of the Americas and Africa effectively isolates the Atlantic Ocean from the Indo-Pacific today, which has resulted in species endemic to each region. The opening of a seaway connecting separate oceans quickly results in mixing of faunas, a process which we have been able to see in action in the short time since the Panama Canal was opened.

In the last 50 years there have been attempts to trace the history of continental distribution back still further. Why should we suppose that plate movements only started with the disruption of Pangaea? It is now certain that Pangaea itself was only a phase in the development of the face of the Earth. The supercontinent was the product of an earlier phase of drifting that brought together other, separate plates, so that in the Ordovician the world had a number of dispersed continents, rather as we do today. Since the process of continental drift continued far back into the Precambrian, it is a difficult business to reconstruct the patterns of very ancient continents, and seas that have long since disappeared into subduction zones. However, the mysteries of these distant times are starting to be unravelled; geophysicists can now identify at least three ancient 'Pangaeas' in the Precambrian, each separated by another phase of dispersed continents. The Earth has been redesigned several times.

MAJOR CYCLES IN EARTH HISTORY: FLUCTUATING OCEANS

Fossils of marine animals can now be collected from rocks covering the deep interior areas of continents that have been stable blocks since the Precambrian, regions which are now far removed from the oceans. In some cases marine deposits of this kind are explained by the fact that regions now isolated from the open ocean were formerly at the edge of continents when they were separate drifting plates. The Himalayas, the origin of which was described above, are an example where sediments that accumulated at the edge of the drifting Indian continent, and Asia, were sandwiched between the two continental blocks as they collided. There is no problem in explaining the presence of thick marine deposits here, the former sites of seaboards that have been crushed during continental collision. In other cases, such as the interior of North America or Australia, there have not been comparable continental collisions and yet the sea evidently flooded over the continental interior from time to time, leaving deposits containing marine fossils extending deep into the heart of the continents. The only explanation for these kinds of deposits is that the sea has periodically extended much further over the continental areas than it does today.

LEFT: Flat-lying marine sediments like some of these in the Grand Canyon in the interior of North America indicate that in the past the sea extended over the continent.

These periods of flooding (or marine transgressions) are now recognised as one of the most important cycles of physical events that have affected the Earth, and their effects on the course of evolution have only recently been considered. Major marine transgressions of this kind have occurred at intervals throughout geological time. Since they are due to the influence of rising sea level, they are simultaneous over the whole world. They afford another way of subdividing geological time. When the sea extended over the widest areas, naturally the deposits of that age are the most widespread and tend to be the most well known. Conversely, there were periods when the sea drained off the continents (regressions); at these times marine deposits were confined to areas peripheral to the continents and in the open oceans, while terrestrial sediments extended over the areas where marine deposits had accumulated during the transgressive phase.

These oscillations may be related to continental movement. Some scientists have suggested that the periods of transgression correspond with active phases of generation of new ocean floor at the mid-ocean ridges, and the regressive phases correspond with periods of temporary standstill of drifting activity. Another cause that has been invoked is ice ages. During a time of ice advance an enormous quantity of water is locked up in ice-sheets, and world sea level falls as a result. As the ice melts, sea level rises, producing a transgression. Rising and falling sea levels in the last 1 million years of the Pleistocene Ice Age were certainly controlled by glacial events. Whatever the cause, these great cycles have influenced the spread and evolution of life, both on land and in the sea. During the transgressive phase shallow marine faunas are widespread and diverse; in tropical waters reefs build up, and support some of the most diverse communities of marine animals. Sudden changes in conditions were one of the factors leading to marine extinctions.

BELOW: This specimen, known as tillite, is the consolidated remains of the rock fragments and powder that are left behind as a glacier melts and retreats. This tillite from Kimberley, West Griqualand, South Africa provides evidence of much cooler conditions on a continent which today has a climate too warm for glaciers to form.

MAJOR CYCLES IN EARTH HISTORY: FLUCTUATING CLIMATES

As we have seen, the continents are constantly on the move, and the seas may advance upon them and drain off again. This may already seem a world in which everything is in a state of flux, but to provide the environment in which past organisms lived we have to introduce one more cycle of change. The world climate has varied, from generally warm to glacial phases. This has to be distinguished from changes of climate on any one continent; if a continent is moving under the influence of plate tectonics it may pass across climatic belts, starting in the tropics and finishing up near the pole (like the southward-drifting Antarctica). Climatic cycles involve climate change over the Earth as a whole. Such changes are related to global sea level changes, as the major glacial episodes will result in a regressive (draining) cycle over the continents.

There have been a number of major periods of glacial activity in the period over which life has left its abundant traces in the rocks. An important event in the Late Precambrian 600 million years

Glacial ice

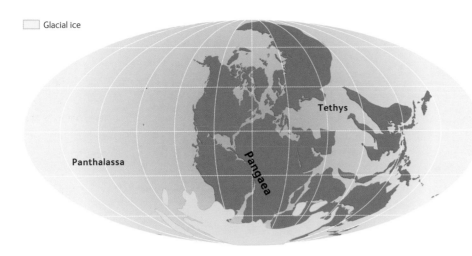

LEFT: The extent of the
Carboniferous-Permian glaciation
in Pangaea, around 280 million
years ago

ago) was probably preceded by others in the further reaches of Precambrian time. This
glaciation preceded the major diversification of animals with hard parts at the base of the
Cambrian period 542 million years ago. According to some scientists this was the biggest
glaciation ever, producing what has been described as 'Snowball Earth'. It is certainly true that
glacial influence extended all the way to the tropics at this time, but whether the Earth was
entirely frozen is debatable. The same scientists link the end of the frigid conditions with the
subsequent explosion of life. Another important glacial phase occurred late in the Ordovician
when a major proportion of marine life was exterminated. The third, and probably the most
important glacial interval (or 'glaciation') was the Carboniferous–Permian event, or rather

LEFT: Glacial striations left by the
recent Ice Age on Precambrian
rocks polished by glacial action,
Churchill, Manitoba, Canada.

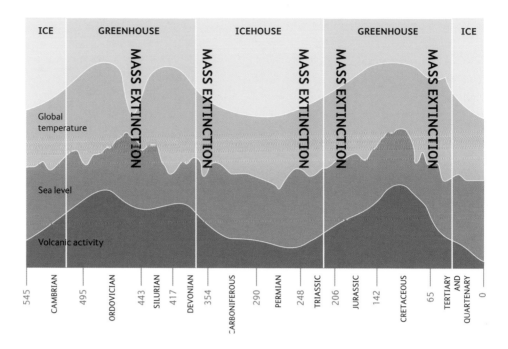

RIGHT: Changes in volcanic activity, climate and sea level influence each other and may cause mass extinctions.

events, the effects of which are preserved over a huge area of the southern hemisphere. The distribution of the glacial rocks here was used by Wegener in 1915 as one of the arguments in favour of 'continental drift'. The effects of this glaciation in the southern hemisphere were also felt in Carboniferous Europe, which at that time lay near the equator. Glacial retreats released immense quantities of water, which resulted in sea level rise. On some occasions the rise was sufficient to drown the coal measure swamps, inundating what had formerly been luxuriant jungle, and interpolating marine fossils in terrestrial rock successions. This Late Palaeozoic glaciation lasted for a long time; between the major pulses of glaciation in the Carboniferous and the Mid-Permian there was an interval of approximately 20 million years.

We are currently at the end of the last glacial phase: the Pleistocene period. Radiometric dating techniques, combined with careful analyses of fossil pollens and various microfossils, have enabled a subdivision of geological time for the last million years on a much finer scale than for the earlier parts of the stratigraphic column. The story so revealed is very complex. The ice has advanced and retreated numerous times; the original idea of four great advances has been shown to be a gross underestimate. The number of glacial phases is now thought to be approximately 15. Phases of retreat between ice advances included times when the climate was very much milder than today. There is no special reason to suppose that the process is at an end; another ice advance may be on the way in the future. However, man-made global warming may drive the climate back to what it was well before the last ice age. The huge quantities of water locked in the present ice-sheets mean that the area of exposed continent is probably greater now than it has been for most of the Tertiary – but that could well go into reverse as climate change begins to 'bite'.

Between glacial phases there were other periods in which the climate of the world, as recorded in the rocks, was generally warmer and more equable. For example, the Silurian and much of the Cretaceous were times of unusual warmth. At these times the tropical zones expanded to cover a greater area of the oceans and continents. Wide spreads of warm-water limestones are typical of the rocks deposited at such times, and with the limestones are found fossil animals adapted to the warm oceans. The climatic changes themselves are partly under the control of atmospheric conditions. The warmest phases tend to be 'greenhouse worlds' when greenhouse gases, especially carbon dioxide, reach high levels. 'Icehouse worlds' occur in times when carbon dioxide levels are low. Now that we are moving into a greenhouse world induced by our own excessive burning of fossil fuels, the outcome may well control the future of our species.

THE CHANGING WORLD

The arrangement of the continents changes, the sea levels change, and the world climate changes. This is the shifting stage on which organic evolution has acted and will continue to do so. The cast of characters has changed repeatedly, the animals or plants adapting to the conditions pertaining while they lived. In this dynamic world, what are we to make of Charles Lyell's ambition to interpret the geological past by processes that operate at the present day? The basic assumption still stands: though the stage may have changed, the processes that shaped the scenery were the same in the past as they are at the present. Physical laws do not change. It is only by a thorough understanding of the forces at work today that the past can be reconstructed. Life is bound up with the story of the changing Earth, and it is foolish to pretend that the history of life can be fully understood without its dynamic setting. The history of life has been so closely bound up with the history of our planet that it is likely that some small change in that history would have produced a change in the course of evolution. If the climate had not changed in Africa a few tens of thousands of years ago, would modern humans have evolved? There is nothing inexorable about the course of evolution; rather it is a complex *pas de deux* between the changing environment and the capacity of organisms to respond to those changes.

Rocks and fossils

A visitor to a museum will see perfectly preserved and spectacular fossils neatly arranged in glass cases. It is easy to forget that these specimens were found by cracking open rocks. All fossils are found in rocks that were originally unconsolidated sediments, and the study of fossils can be enhanced by knowing about the rocks that enclose them.

RIGHT: The Aberystwyth Grits in Ceredigion, Wales is a thick formation of Early Silurian rocks (443–428 million years old) consisting of deep-water shales, alternating with coarser beds deposited by sudden turbidity currents.

Certain environments which today support a rich and varied plant and animal life have no sediments forming in them, and the organisms living there have virtually no chance of being preserved in the fossil record. Mountainous regions, for example, are dominated by the erosion of the rock forming the ranges, and therefore no permanent sediment is formed there. Torrential rain and rapid weathering, aided in some climates by the action of frost, rapidly destroy much of the organic material, and the chances of any preservable remains reaching a lowland river where permanent sediment is accumulating are remote. As the fossilisation potential of a mountainous environment is low, the faunas and floras of mountainous regions of the past are most unlikely to be represented in the fossil record.

The study of fossils is connected with a suite of rocks that are formed in environments where sediments accumulate, and have a high chance of becoming rocks. Such environments cover a large part of the surface of the globe, including most of the submarine areas, and some of the lowlands, where rivers and lakes accumulate sediments of many types. This chapter is concerned with the different sites in which sediments may form.

In some instances the site of sediment accumulation is a direct reflection of the environment in which an animal or plant lived. Lake fishes and plants, for example, are to be expected in lacustrine (deposited in a lake) sediments. The wide variety of sediment sites will be reflected in an equal variety of animals from different habitats. Extensive studies of recent sediments enable the interpretation of sediments from the past. Most sedimentary rocks retain in fine detail the features acquired while they were accumulating. By studying recent sediments it is possible to determine the site of deposition of past rocks, and from this to understand more about the environment in which the fossil animals in the rock lived.

Occasionally dead animals and plants travel for long distances before finally becoming entombed in the sediment. Empty shells of *Nautilus* have been found over a much wider area than that in which the animal lives. Drifting logs can be found hundreds of kilometres from land, and when these become waterlogged, they sink and eventually become incorporated into the sediment.

SEDIMENTARY FACIES

The rocks formed at a particular site, each with their own peculiar characteristics, are called sedimentary facies. Just as at the present day, sediments of many different facies are accumulating in different places, so in the past rocks with totally different appearances may have accumulated at the same time in different environments. The fossils found in such rocks also differ from one site to another, as they are related to the environment in which the sediment accumulates; different sedimentary facies may have different assemblages of fossils. The term facies fauna is applied to an assemblage of different fossils that are found together in one particular sediment type. Not all animals are so limited: in the sea, for example, many free-swimming or floating organisms are independent of the conditions of sediment accumulation on the sea floor.

The diagram below shows the main sedimentary facies in a hypothetical section running from the mountainous interior of a continent in tropical latitudes to the open ocean. In the lee of the mountain range a rain-starved desert accumulates mostly wind-blown sand derived from the weathering of steep buttes. Occasional torrential bursts of rain produce flash floods, which sweep down the steep-walled valleys (or wadis) carrying with them all the weathered

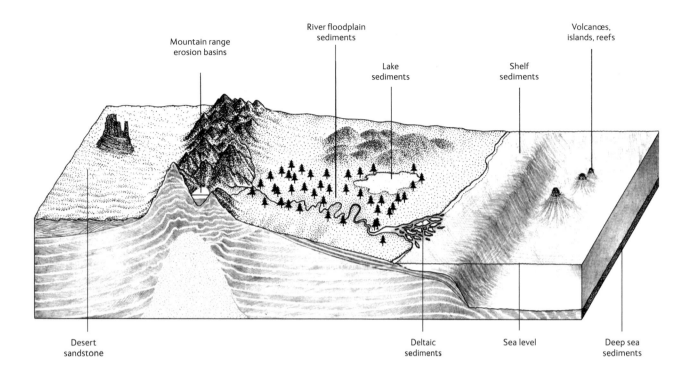

Mountain range
erosion basins

River floodplain
sediments

Lake
sediments

Shelf
sediments

Volcanoes,
islands, reefs

Desert
sandstone

Deltaic
sediments

Sea level

Deep sea
sediments

material, which spreads out into broad fans. The floodwater drains into temporary pools, which evaporate rapidly in the hot sun, but sometimes linger long enough to allow brief bursts of specialised animals to colonise the warm water. As the waters evaporate, mineral salts are concentrated within the pools, often crystallising out as the drying continues to produce white, glistening saltpans. Most of the animal and plant life here has to be very specialised to cope with the harsh conditions.

On the mountains themselves erosion predominates, aided by the action of ice found on the higher peaks. Frost shattering helps to splinter rocks into shards and blocks that tumble down slopes and become the raw material for streams to move inevitably downwards towards the sea. Melting snow produces raging torrents with immense transporting power, which in full spate can move and break huge blocks of rock into smaller cobbles that are more easily transported. Little sediment accumulates here except in deep depressions between ranges (inter-montane basins), the scree slopes at the edge of the mountains, and in the deposits of streams and rivers. Much of the sediment produced on the mountain sides becomes ground into smaller particles. In this state it can be transported by the large and more sluggish rivers through the foothills and beyond. Occasional floods originating in the mountains may transport huge amounts of material, often with disastrous consequences such as flooding on the plain.

The deposits associated with major rivers are silts, clays and sands, often with characteristic combinations of sedimentary structures that reveal their fluviatile origins. Large lakes lying on the plains are important and potentially fossil-bearing sites of sediment accumulation. In low-lying areas swamps support prolific vegetation, which decays to form beds of peat. Insects and other animals adapted to this habitat may be destined to be preserved therein.

ABOVE: A section from a desert area through a mountain range, across a sedimentary plain to the shallow sea, and ultimately to the deep sea, showing the different sites in which fossils may accumulate.

As the rivers wind towards the sea, their floodplains broaden and they meander over the plains, breaking their banks and readjusting their courses at times of flooding. They can carry huge quantities of sediment, mostly in suspension (at the present day about 10,000 million tonnes of sediment finds its way into the seas over the whole world each year) but also in solution. At the junction of land and sea the river begins to shed its load of sediment, partly from the effect of freshwater meeting salt water, and partly because its energy and carrying power dwindles. Typically the river builds a delta, and the silts, sands and clays of deltas are one of the most important sedimentary facies in the fossil record. Swamps may form around the small streams (distributaries) that criss-cross the delta, and there are many habitats suitable for the successful life of animals and plants, as well as their rapid preservation on death. Deltas slowly build out into the sea, forming an advancing wedge of sediment that may extend for many kilometres. In such cases the seaward edge of the delta is younger than its fossilised edges preserved on the landward side. Deltaic sediments, although in geographic continuity, are not all the same age. These are diachronous deposits. Smaller rivers, like the River Thames, often do not build deltas but enter the sea in estuaries, where salt water and freshwater oscillate in influence in tidal stretches of water. Estuaries also produce characteristic sediments, and have specific sets of organisms associated with them.

We now pass into the marine environment, which has produced the bulk of the sedimentary rocks and in which the record of past life is most complete. In many areas the contact between land and sea is erosional, as any visit during a storm to a resort with sea cliffs will prove. Storm beaches of rounded cobbles often form the boundary between the sea and the land, and may be preserved in the rocks as coarse conglomerates. Further

ABOVE: Fossil ripple marks preserved in sandstone. The detailed structure of the ripple marks can reveal much about the past sedimentary environment.

RIGHT: An oolitic limestone from Rutland, England, showing the perfectly rounded ooliths of which it is composed. Such limestones formed in shallow, agitated marine conditions, and only in warm climates.

offshore, sediments are generally finer sands, silts and clays, and far more suitable for preserving fossil remains than the deposits of storm beaches. The finer material from rivers and from the direct erosion of the land by the sea contributes to the sediment deposited on the sea bed. Powerful currents on the sea floor often influence the pattern of distribution of such sediments. Shallow marine deposits also have characteristic sedimentary features; the ripple marks of tidal seas are frequently encountered in rocks, and the tracks of animals can be preserved here. Fossil footprints in sand are no doubt preserved somewhere today, recording the holiday habits of humankind. Although there are many exceptions, the grain size of the sediment generally decreases away from the coast, so that deposits further offshore tend to be fine muds that will produce clays or mudstones in the geological column.

In warm climates, where there is not much land-derived sediment available, other forms of sedimentation become important, especially the formation of limestones. Many limestones are formed from the consolidation of lime muds, the precipitation of which is mediated by bacteria. The lime in these muds is a form of calcium carbonate (aragonite) with minute crystals, which can only be deposited in warm and shallow seas. Where there is turbulent agitation of the water in shallow areas, the carbonate may be laid down in concentric layers around tiny organic nuclei to produce spherical ooliths. A rock largely composed of ooliths is an oolite. In spite of their rather specialised mode of formation oolites can form a surprising volume of limestones, covering hundreds of square kilometres in the Ordovician rocks of North America, and almost as extensive in the Carboniferous and Jurassic rocks of Europe. Numerous species of animals are confined to a lime substrate, and naturally one finds their extinct counterparts preserved in limestones. Some limestones are composed almost entirely of the remains of calcareous animals.

As the deep sea is approached there is generally a reduction in the amount of sediment that can reach the open ocean from land sources. On the floor of the open ocean at great depths, large areas are covered with a fine ooze, which is formed predominantly from the skeletons of minute planktonic organisms that have rained down from the surface waters. The greatest density of planktonic life exists in shallow depths where light penetrates and allows microscopic plants to live, as well as small organisms which feed on these plants; it is the shells of these small animals that form much of the deep-sea sediment. The rate of accumulation of these deep-sea deposits is very slow, only a few millimetres per thousand years. Several kinds of single-celled organisms may dominate these deep-sea oozes; the foraminiferan *Globigerinoides*, which has a calcareous test,

BELOW: The foraminifera *Globigerinoides*, which can be dominant in deep-sea oozes. The specimen is 2mm (1/12 in) across.

and the delicate, siliceous radiolarians are especially important. These deep-sea deposits record a remarkably complete history of the evolution of the planktonic organisms contained within them. Even these small shells, however, are to some extent soluble in sea water, and at very great depths pressure increases this solubility so that they do not survive. Here the only deposit is red clay (often in fact brown in colour), a sediment which accumulates extremely slowly. It is composed of the finest wind-blown dust, volcanic ash carried by winds from distant eruptions, and, occasionally, the insoluble traces of animals such as the teeth of sharks or the ear bones of whales. In these abyssal depths curious nodules with a high proportion of manganese grow slowly in the red clay, and there has recently been speculation about the possibility of exploiting these as a mineral resource, surely the least accessible ore in the world. In spite of the inhospitable, light-less conditions in the abyssal seas there is a variety of life, and there are specialised and often bizarre fish and crustaceans that live only there. These leave little fossil record.

In the open ocean, volcanic islands form sporadic sediment sources, both from the erosion and redistribution of the volcanic rocks themselves, and because they reproduce the same sort of conditions that pertain on the continental shelves. The clear water surrounding such islands in the tropics is often suitable for the growth of coral reefs, discussed later in this chapter.

In Arctic regions the influence of ice as both an erosional and depositional agent is paramount. The scouring action of ice, sometimes using rocks enclosed within glaciers as tools to scrape and gouge the underlying rock surfaces, produces great quantities of angular detritus. Some of the rock is ground as fine as flour. At the melting edges of glaciers, or where icebergs break off from ice-caps and drift into the sea, much of the material is released and becomes sediment. Such glacial deposits (till) are often a heterogeneous selection of different rock types, dumped together, with large boulders and tiny pebbles immersed alike in the groundmass of rock flour. Not surprisingly fossils are rare in these kinds of rocks. Around the edges of ice-sheets mossy bogs are common and may form deposits of peat and lignite containing the remains of organisms adapted to life in high latitudes. During the last ice age animals often used caves as shelters or lairs, and the deposits of cave floors have proved particularly rich sources of their bones. Further away from the ice front, major rivers took away the meltwaters to the sea, and their fluviatile deposits often preserve the remains of large mammals that lived in the surrounding areas, some no doubt fatally entrapped in bogs.

Some land-derived sediments reach the deep sea by means of turbidity currents. These are slurries of sedimentary material that are flushed from shallower areas at the edge of the continental shelf, a movement often sparked off by earthquakes. Sometimes their effects can be quite catastrophic, snapping underwater cables, and they can travel extraordinary distances, up to 300 kilometres (190 miles) or more. Turbidity currents produce a characteristic rock type in the geological record known as a turbidite. Communities of animals that live on the sea floor and suddenly have material deposited on top of them by a turbidity current can be buried in a single catastrophe.

Just as climate influences the kinds of sediment laid down, so also it is one of the most important influences on animal and plant life. Few species are truly global. Most marine organisms are distributed according to latitude, and it is possible to represent these distributions as a series of belts approximately parallel to lines of latitude (distorted by the

OPPOSITE: The delicate, siliceous skeletons of radiolarians such as *Acanthometra* with skeletal spicules, cover large areas of the deep-sea floor (here with foraminifera, lower left).

RIGHT: Climatic fluctuations in the Pleistocene are often recorded in cave deposits. These bone stacks built by William Beard at Banwell Bone Cavern, UK, are nearly all those of the bison *Bison priscus* and are about 70,000 years old, indicating very different climatic conditions to today.

influence of warm and cold currents). The same sort of influences undoubtedly operated in the past. During the Pleistocene, when climates oscillated over many thousands of years between warm and cold, marine and land organisms migrated backwards and forwards with the climatic shifts to keep living in the conditions to which they were adapted. Since these oscillations ran approximately parallel on land and in the sea, this provides one of the methods of subdividing the ice ages. Tropical faunas and floras are richest in numbers of species, and, at the other extreme, only a few hardy species of high Arctic or Antarctic animals and plants are able to cope with extreme polar conditions. Those species that do adapt to polar conditions may be found in great profusion.

We have seen how sediment type alters with the depth of water. Animals are also influenced by water depth, and it is not surprising to find different communities of fossils in rocks that were deposited at different depths. Both the sediments and the animals found as fossils within them are reflections of the past environment, and both provide clues to the life and conditions of former times.

BELOW: The global distribution of plankton (graptolites, left image and crustaceans, right image), showing how these organisms have a latitude-controlled distribution.

Ordovician 495 million years ago

Present day

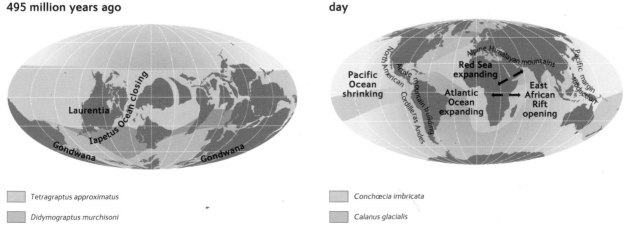

| Tetragraptus approximatus |
| Didymograptus murchisoni |

| Conchœcia imbricata |
| Calanus glacialis |

REEFS

Organic reefs today are formed by corals and calcareous algae that make wave-resistant structures, often of immense dimensions. Because the framework of a reef is composed of tough calcareous organisms that have to stand up to the buffeting of waves and the ravages of storms, they are likely to be preserved in the fossil record. Reefs can be preserved with all the constituent organisms in life position. Living coral reefs can be found in warm water regions (23–25°C or 73–77°F is best) with abundant sunlight. Light is needed by the minute photosynthetic algae that live in the tissues of some reef-building corals. Reef growth can only proceed satisfactorily in salt water, and not in water of too great a depth where the vital light begins to be filtered out. Reefs cannot tolerate too much clogging sediment in the surrounding waters. The majority of fossil reefs seem to have lived under similar constraints. However, in the present-day North Atlantic Ocean some cold water corals that do not need photosynthetic algae form reef-like structures down to almost 1,000 metres (3,300 feet) water depth.

The reef environment is an extremely rich one, supporting a host of animal species besides the frame-builders themselves: fish, molluscs, sea urchins and crustaceans live among the corals as scavengers and predators, together with numerous encrusting organisms that benefit from the firm foothold provided by dead coral. The corals themselves feed by filtering minute planktonic food from the constant wash over their polyps. Erosion of the reef is rapid, and this builds slopes of reef waste and finer debris that can also be found fossilised. This produces a rock that is almost entirely made of fossils. Charles Darwin was the first to demonstrate that the curious circular pattern of atoll reefs was due to the foundering of volcanic islands, around which reefs had initially formed as fringing structures. The rapid growth of the reef can keep pace with the sinking of the island, producing in the process a great amount of sedimentary waste that preserves the earlier, fossil history of the reef.

The reef as a structure has a long history, extending back even to a time before the corals themselves had evolved. Other organisms were able to adopt the same role as the corals, often assuming similar general shapes, even though unrelated in a zoological sense. The

LEFT: Corals forming on a reef at Kimbe Bay, Walindi, Papua New Guinea.

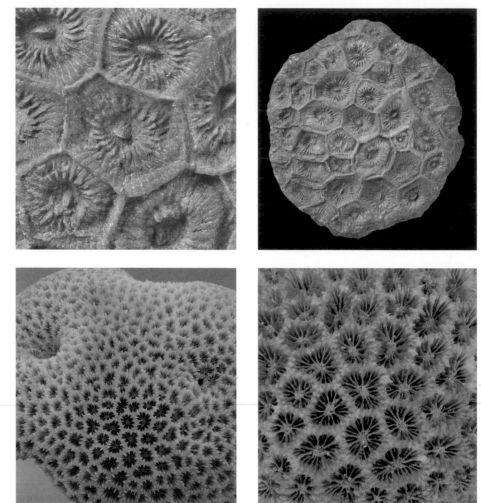

TOP LEFT AND RIGHT: Fossil coral *Actinocyathus floriformis* (order Rugosa) from Carboniferous coral rich beds, Shropshire, England.

BOTTOM LEFT AND RIGHT: Recent coral *Favites pentagona* (order Scleractinia) from a coral reef along the coast of Somalia.

earliest reef structures are present in Early Cambrian rocks in Labrador, where the frame was built by sponge-like organisms called archaeocyathids. Corals themselves, although belonging to forms unrelated to living species, started to achieve prominence in the later Ordovician, and by the Devonian large reef structures were present. The storm-facing surface of the reefs was often formed by massive stromatoporoids, a group of organisms with only a few inconspicuous relatives living today. Fossil reefs are found almost worldwide, reflecting the warmer temperatures globally at certain times in the past. Some of the corals from these reefs closely resemble their living counterparts although this resemblance is due to similar life habits, and their detailed structure is quite different.

In the Permian rocks of Texas and adjoining states, large reef structures have been found where sponges and bryozoans were the important frame-builders, along with the ubiquitous algae. The same habitat supported one of the most bizarre brachiopods, a conical form with a 'lid' quite unlike the usual brachiopods. At the end of the Permian most of the reef-building organisms that built large reefs in the Palaeozoic became extinct. By Jurassic times coral reefs were again being constructed, this time by the distant relatives of the corals and other organisms that build reefs at present. The corals of Palaeozoic age are distantly related, if at all, to those of the Mesozoic to Recent, but they both built reefs of similar construction. During the Cretaceous an extraordinary group of molluscs acquired the habit of building sea-floor structures, although they were not really reefs. These were the rudists, which again are conical, with a lid, at first glance much like the aberrant brachiopods of the Permian. However, the rudists were quite unrelated to the brachiopods, and were derived from clam-like molluscan ancestors. The rudists did not survive the Cretaceous 'greenhouse world', and were confined to very warm limestone seas; they may have been adapted to higher temperatures than pertained over much of geological history. They are important rock fossils in the Alps and in North America, where they form masses lying on the ancient sea floor. During the Tertiary, the break-up of the continents, with the associated volcanic activity creating islands in the oceans, permitted the establishment of the ancestral reefs that continue to be built until the present day.

The reef environment is important because it shows how different organisms can assume a similar superficial appearance when they adopt similar life habits. There is more than a passing resemblance between some of the archaeocyathids of the Cambrian and the Permian brachiopods, and some of the corals have comparable forms as well. On present-day reefs more densely branched forms tend to be found on the exposed, seaward flanks, while the backreef areas have organisms with loosely branching antler-like growths. There may be variation in branching habit even within a single species, according to the site in which it flourishes. Through geological time various organisms have played the same ecological role, and the result has been similar shapes. This is an ecological control that can act on very different starting material (like bivalves and brachiopods) and produce a superficially similar end product. The biological prerequisite of most of the organisms mentioned here is that they should be filter-feeders. For a highly adapted organism like a reef dweller it is necessary to look carefully at the fine structure to determine the biological affinities of the organism, and not be misled by superficial resemblance.

BELOW LEFT: The bizarre, specialized bivalve molluscs known as rudists often found in the warm seas of the Cretaceous period.

BELOW: A Cretaceous rudist *Hippurites socialis* from the southern central Pyrennees, Spain. The valves can be 10–30 cm (4–11 3/4 in) high.

DEEP-SEA DEPOSITS

The peculiar sediments formed in the deep sea, at or beyond the edge of the continental shelf, also have a long history. Deposits laid down in the open ocean occupy a large area of the globe today, and there is every reason to suppose that this was true far back in the Palaeozoic. However, the area occupied by oceanic sediments in regions of Palaeozoic rocks is not commensurate with this former extent. This is because most of the ancient oceanic sediments have been destroyed where they plunged down the subduction zones at the consuming edges of plates; oceanic sediments are the ones that disappear forever. The crust of the Earth is in a state of dynamic equilibrium, with oceanic crust being created at mid-ocean ridges, only to be consumed at the edges of plates. The break up of the supercontinent Pangaea during the Mesozoic resulted in both the creation of oceanic crust that floors the oceans today, and the destruction of previous oceanic crust, so that at present the oldest oceanic crust is probably only Jurassic in age.

ABOVE: Agnostid trilobite of Cambrian age, often found in deep water. It is a blind trilobite, only a few millimetres long, with only two thoracic segments.

In spite of the odds against it, some ancient oceanic deposits are preserved, but only in special circumstances. Sequences of turbidites form prisms of sediment, sometimes thousands of metres thick, at the edges of the continents. During phases of collision these sediments become squeezed between the colliding plates as the oceanic crust itself is consumed. The sediments respond to this pressure by crumpling, shattering, and gliding into great sheets that move away from the centre of pressure. Many of these sediments become heated, partially melted, or so contorted by pressure that any fossils they once contained are transformed beyond recognition. Some, however, survive with their fossils intact, although the strata from which they have to be recovered are almost always vertical, rather than in their original horizontal attitude, and frequently heavily distorted, with the fossils they contain distorting along with the rock. The fossils are not usually found in the turbidites themselves, but in interbedded shales, representing the quiescent conditions between turbidite slumps. In some cases slices of oceanic rocks, instead of being consumed, have been thrust beyond the danger zone, often carrying on their backs a skin of sediment. These slices have a characteristic combination of volcanic rocks (often with serpentine) with cherts and sometimes black shales. These remnants of former oceans are known as ophiolites. Where ophiolites are found in mountain belts this indicates that oceanic rocks have been obducted in that region. Adjacent areas of folded or metamorphosed rocks were produced by continent colliding with continent or island arcs docking, much of the intervening ocean which was originally present having dived to oblivion. In order to find deposits laid down in the ancient deep seas and the fossils they contain, the right geological setting must be found. It is of no use looking in the sediments formed on the ancient shield areas of Precambrian rocks, which even in the Palaeozoic were covered by relatively shallow seas. In the areas of intensely folded rocks that probably accumulated at the edge of former continents, the chances of finding deep-sea sediments are increased.

In the Cambrian, deep-sea shales between turbidite sequences have been found to contain trilobites, minute, blind forms, known as agnostids. The likelihood is that these specialised trilobites were free-swimming or planktonic forms, and they did not actually live on the sea floor. In the Ordovician and Silurian drifting colonial organisms known as graptolites usually dominate any assemblage of fossils recovered from between the turbidites, their remains often forming matted clots. Clumps of graptolites, perhaps killed due to a plankton 'bloom' consuming the available oxygen, slowly fell through the water column, eventually to lie down on the soft muddy sea bed; the arrival of another turbidite flush would make their entombment complete. The occasional glass sponge may have actually lived on the deep ocean floor; they are found sporadically through the fossil record in these sorts of sediments, and are still relatively prolific at the bottom of the oceans today. Special extraction techniques can recover the remains of single-celled plants from marine deposits of Palaeozoic (or even Precambrian) age. These show that photosynthetic algal plankton, the basic link in the food chain of past seas, were present from the earliest times. No doubt the seas also swarmed with minute zooplankton feeding off the tinier plants. Such zooplankton were mostly soft-bodied and left hardly any record. The carapaces of some extinct crustaceans are known in abundance in graptolitic deposits; they may have fulfilled the same plankton-feeding function as the open-sea shrimps today, but there was certainly no Early Palaeozoic 'whale' to harvest them. Cherts accompanying ophiolites often contain the remains of microscopic radiolaria, which had acquired their siliceous skeleton even by the Cambrian, and must have had planktonic habits then, as at present.

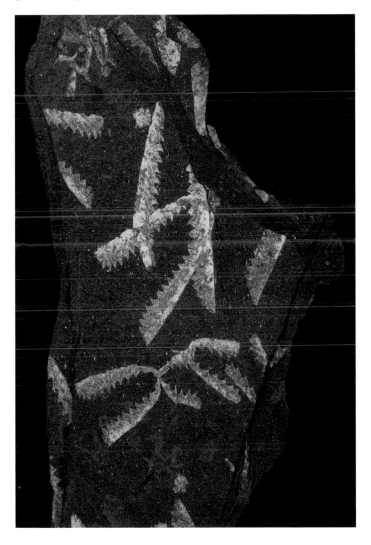

BELOW: Tuning-fork graptolite *Didymograptus murchisoni*, Ordovician, Wales. Several of these graptolites are preserved on the flat bedding surfaces of a black shale. The graptolites are flattened, and their original skeletal material has been destroyed. Individual specimens grow to a length of 5 cm (2 in) or more.

The graptolites became extinct early in the Devonian, and no ocean-going animals of this colonial type are found subsequently as fossils. At about the same time the early relatives of the ammonites were adapting to life in the open seas, and their coiled shells, variously ornamented, become the fossils most frequently encountered in deeper sea sediments for several hundreds of millions of years. Ammonites swarmed in vast numbers, both in the

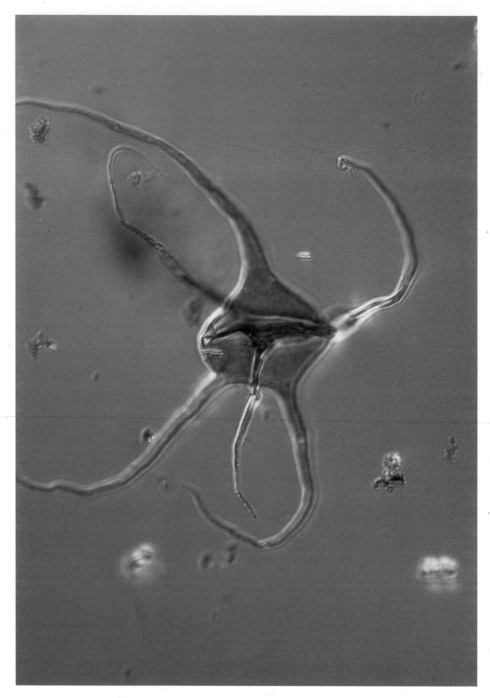

seas of the continental shelves and in the open sea, and the total mass of living matter they must have represented over their geological lifespan from the Devonian to the Cretaceous is almost inconceivable. Some species may have swum in masses, moving gregariously like their distant relatives the squid, which occur in huge numbers in the open seas today. By the later Palaeozoic fierce predators added another level to the food chain, the sharks by then being well advanced, and doubtless as voracious as they are in modern seas. The sharks are an ancient group of fish in their structure and organisation, but most of them are superbly well adapted to their role as swimmers and hunters, which has ensured their survival into modern times, while their early prey species, such as ammonites, have passed into extinction. Along with ammonites other kinds of invertebrates, like delicate clams, sometimes occur in profusion in Palaeozoic and Mesozoic rocks. These may either have been free-swimming forms, or attached to floating seaweed. The microfossils include an increasing variety of species of planktonic algae, which by the Jurassic included forms related to those still living today. The radiolaria continued their unbroken history from the earliest fossil-bearing rocks. In the Jurassic the foraminifera, hitherto almost exclusively bottom-living forms, took to life in the surface waters of the oceans, and played an increasing role as sediment makers, which continued until the present day. During the Cretaceous minute calcareous platelets which

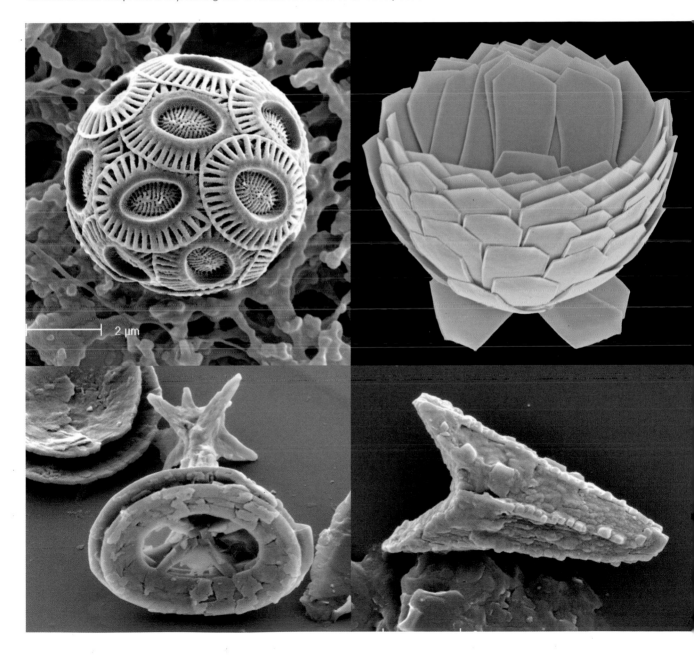

coated the outside of algae (coccoliths) were an important component of pelagic sediments, and for all their small size show a wonderful variation in symmetry. The slow build-up in the variety of animals contributing to oceanic sediments continued until the Late Cretaceous, when, apparently quite rapidly, the ammonites became extinct.

Whatever the cause of the demise of the ammonites, it did not affect the foraminifera or coccoliths to quite the same extent, although the post-Cretaceous foraminifera are nearly all new forms. For Cretaceous and younger rocks we can investigate the deposits of the deep seas directly by taking cores from the bottom of the sea, or looking at rocks where they have been uplifted on volcanic islands. During the Tertiary the oceans acquired a modern aspect, with fish, whales, modern crustaceans, and micro-organisms related to those of present seas. Nonetheless, a succession of different species can be recognised, with certain types becoming extinct, and others appearing as 'landmarks' for the divisions of the last 70 million years.

The sheer volume of the oceans makes the oceanic environment relatively stable compared with the terrestrial one. Extreme changes of temperature, for example, are muffled in the sea, and below a certain depth the temperature is virtually the same the world over. Particularly in the open ocean the animals are cushioned from the effects of environmental fluctuations that have provoked the rapid evolution of terrestrial organisms. Comfort usually stimulates laziness, but we have seen that the oceanic environment has, in fact, changed considerably since the time in which it can be first recognised in the rocks. The radiolaria and planktonic algae have been with us all the time, but have not remained static. Successful groups, like the graptolites, have failed entirely, while the planktonic foraminifera have successfully adapted to the challenges of the open ocean. It is now known that there were certain periods when the oceans did undergo crisis. These times often entailed the increase and spread of sea water lacking oxygen, and were unfavourable to virtually all forms of life. No doubt these crisis periods were implicated in the changes that occurred in the composition of oceanic fauna and flora. However, it can be shown that the changes that occurred on land, for example among the mammals in the Early Tertiary, did take place more rapidly than those detectable among the planktonic foraminifera. Nonetheless, the cumulative and inexorable changes have resulted in a complete transformation of oceanic life over 600 Ma.

NICHES NOT REFLECTED IN THE ROCKS

The sediment surface indiscriminately receives the scraps that will become fossils, and it would be a mistake to assume that any animal when alive necessarily lived where its remains finish up. Any one environment is usually subdivided into innumerable micro-habitats. The variety of animals living together is explained by the number of niches (particular ways of earning a livelihood) into which a broader environment can be divided. Many of the details of the niche are not reflected directly in the rocks, and therefore the details of the life habits of the fossil plants or animals have to be inferred from different evidence. There are cases where an animal's livelihood depends on numerous kinds of soft-bodied organisms, which have left virtually no record. Polychaete worms are an abundant source of food in shallow marine environments, but their only geological legacies in most rocks are slight disturbances of the sediment produced by their burrowing activities.

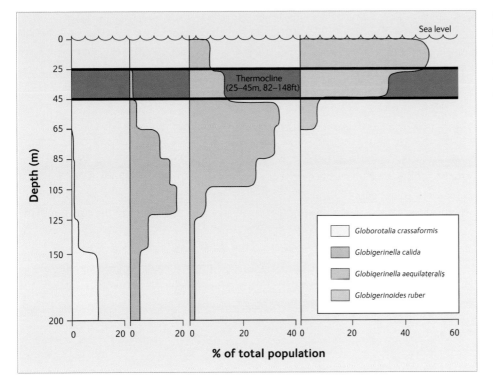

LEFT: Depth stratification to show that different species of foraminifera live at different depths. The sea floor receives dead remains regardless of the depth at which the animal may have originally lived.

Among living planktonic organisms there is often a stratification according to water depth; most species live near the surface, but there are some that live at deeper levels in the water column. The sediment surface below receives both kinds with equal ease when they die. Depth stratification of this kind has been suggested for fossil foraminifera and graptolites. In this case the record in the rocks can be used to test the theory. Shallow water planktonic species will have a distribution through much of the world in the surface waters, but the deep species will be restricted to areas where the water depth is sufficient. The deepest basins should have the richest assemblages of species in the sediments, because it is in such areas that all the depth zones will be stacked up above the sea floor.

Recognising fossils

Fossils are not rare. We have seen how millions of fossils may make up the rocks themselves, crammed together layer after layer to build formations hundreds or even thousands of metres thick. The present land surface is a thin skin on top of a thick record of the past preserved in the rocks.

RIGHT: The Eocene bony fish, *Pristigenys substriatus*, Monte Bolca, Verona, Italy. It has a deep body, narrowly compressed from side to side, very large eyes, prominent fins and a fan-like tail. Length 10 cm (4 in).

Once scientists began to collect the record of past life they soon came across the problem of how to order and arrange the huge variety of fossil forms they recovered. In order to understand what they had found they needed to classify the fossil organisms into particular kinds, more or less similar, to impose an order on what would otherwise have been a vast and chaotic mass of different and apparently unrelated relics.

The urge to order the natural world appears to be an innate characteristic of human beings: "a place for everything and everything in its place". But classifying animals and plants, living and fossil, is more than just an attempt to find a convenient way of slotting them into different categories, like stamps in a stamp album, for neatness and convenience. The classification of the natural world is supposed to reflect the great ordering process that itself gave rise to the variety and diversity of animals and plants that are alive today: the process of evolution. Animals that look similar are classified together, and not only that, they are also closely related in an evolutionary sense (or put another way, they share a common ancestor). So, when we look at the grimacing gestures of a chimpanzee and wonder at the almost ludicrous parallels with our own behaviour, this is just part of a whole host of behavioural and anatomical similarities that show without doubt that we ought to be classified with the apes (we are all of us primates), and that we share a distant ancestor with our diminutive caricatures. Evolution has driven humans and chimpanzee further apart from our shared ancestral species, and in evolutionary terms the chimpanzee is as advanced as we are, although we dominate by virtue of numbers and adaptability. Few people are offended today, as they were in the last century, by the thought of humans and the chimpanzee being classified together by virtue of a common ancestor that has been extinct for probably more than 3 million years!

A word of caution here: not all similarities indicate relationships with evolutionary significance. Some similarities can be misleading, because animals can superficially resemble one another that are not closely related in an evolutionary sense. Often these resemblances

RIGHT: Some palaeontologists think that the richness of the marine environment was established as early as 444 million years ago, but of course the kinds of organisms have changed. This ammonite, now extinct, floats in a Cretaceous sea above crinoids, sea urchins, brachiopods and molluscs.

are the product of a similar mode of life. Both tortoises and armadillos are animals that carry around their own suit of armour, and at first glance we might think they were related. But a little further investigation shows that tortoises are cold-blooded reptiles that lay eggs, while armadillos are in many respects typical warm-blooded mammals, bearing their young alive. Obviously tortoises and armadillos cannot be classified together in spite of their similarities. When classifying fossils even more care is necessary because we have not got as much evidence to go on as with living animals. We cannot see directly whether a fossil animal was warm or cold-blooded when alive; often the evidence is indirect and depends on careful study of little bits of bone. The palaeontologist is like a detective trying to reconstruct a full story from a few fragmentary clues. He or she must be careful not to follow any 'red herrings' that will result in classification of the fossil on the basis of ambiguous similarities. Palaeontologists must distinguish snakes from eels, tortoises from armadillos, on the basis of the bones presented to them. To do this, they must have thorough familiarity with the living fauna, because the information on living animals is so much more complete. We shall see in the next chapter how carrying comparisons with living animals too far can result in curious and inaccurate pictures of the past.

The number of species of animals and plants living today runs into millions, and similar numbers of species have probably lived on Earth for at least part of the planet's history. Perhaps it is fortunate that the fossil record preserved only a fraction of the truly stupendous total number of species that must have lived since the Cambrian, for otherwise, the scientists' task to catalogue billions of years of life would be an impossible one. It is not surprising that new fossil species are discovered daily, and indeed amateur collectors have a good chance of finding a new species of fossil, if they look hard enough and learn to recognise what they have found.

There must have been an increase in the number of different kinds of animals and plants since the Precambrian; for example, the conquering of land alone gave rise to a multitude of new opportunities for the colonising organisms, resulting in an increase in the total number of species. Within the marine environment itself some palaeontologists believe that the overall richness of the marine fauna was established early on, say by the Silurian period, and that there has not been a great increase in the total number of species living in the sea at any one time since then, although of course the kinds of organisms inhabiting the area have changed many times. And there have been several major episodes of extinction, when the cast of characters in the sea changed almost totally. Although the number of species may have been at least approximately the same in marine environments for the last 300 million years or so, the kinds or fossils have changed repeatedly, so that, for example, in marine limestones of Silurian age the shelled brachiopods may number dozens of species, whereas in similar looking limestones of Eocene age no brachiopods at all can be found, but there may be as many species of gastropods of kinds unknown in Silurian rocks.

Even by the beginning of the Cambrian period, when fossils start to become easy to find and many different kinds of animals had acquired preservable hard parts, it is possible to classify the fossils found in the rocks into broadly similar groups. Nearly all of these groups correspond with major divisions of living animals, so the broad base of classification was established by the Cambrian. These major groups continued to evolve throughout the following 500 million years, and they include the distant ancestors of our living fauna. Not all the major living groups of organisms were present in the Cambrian, however, because the colonisation of the land did not take place for another 150 million years, and so there were obviously no direct ancestors of land plants around, or of most of the land animals with which we are familiar today. The major groups into which we can fit almost all fossils from the Cambrian onwards, and all the living fauna, are called phyla (singular: phylum). To identify any animal properly, the first stage is to determine to which phylum it belongs. Of course even the phyla themselves originated from unknown ancestors, and they all ultimately derived from the first living cell. Fossil evidence for some of this early history is shrouded in obscurity, hidden in the vast stretches of Precambrian time, but as is shown later more evidence is being discovered each year. Some of the answers are only now coming to light as we find out more about the structures of the proteins that go to make up living cells themselves. By the time organisms had become sophisticated enough to have hard parts, evolution had already defined the phyla, and the great natural framework for classification had been almost completely constructed.

At the opposite end of the scale from the phylum is the smallest unit of classification usually used for fossils: the species. To be strictly accurate a species can be defined precisely only in living animals, where it refers to populations that can interbreed under natural conditions, and which produce offspring that are capable of further reproducing their kind. In practice, many species are recognised by having some peculiarity of shape, behaviour, plumage, colour and so on, which reliably sets them apart from all other similar species. Obviously we shall never know whether fossil species were capable of interbreeding until somebody invents a time machine, and we can go back and see for ourselves. So our fossil species are defined in a rather practical way, as showing some consistent difference or differences from all other related species. Because of the patchy nature of the fossil record many species (particularly of large vertebrate animals) are known from only one specimen; some of these are of almost inestimable value, which is why museums have to take good care of their material. Every species has a unique scientific name. The species are 'christened' when a scientist describes them for the first time, illustrates their peculiarities and publishes the name in a scientific journal.

The species name has two words: *Tyrannosaurus rex* is a familiar example. The names are usually derived from Latin or Greek – 'Tyrannosaurus' means 'tyrant lizard' and 'rex' 'king' so the animal is appropriately described as King of the tyrant lizards: *rex* is the specific name, unique to this species. There may be other dinosaurs, which are similar to *Tyrannosaurus rex*, but belong to another species. These may be included in the same genus – *Tyrannosaurus* – but will be given a different specific name. Some species are named after their collectors, so if our new species of *Tyrannosaurus* were collected by a Mr Jones it might eventually be christened *Tyrannosaurus jonesi*. Incidentally, the generic name always starts with a capital letter, and the specific name with a small one, even if it is named after Jones.

In this chapter, typical examples of the kinds of fossils most commonly encountered are illustrated by beautiful specimens. Very few fossils have everyday names, and so the scientific name is used. Once a few have been mastered it is surprising how quickly the most ponderous sounding scientific name acquires a familiar ring. The scientific name has the great advantage of being the same all over the world, so the language of nomenclature is a truly international one.

Between the species on the one hand and the phylum on the other there are a whole series of intermediate categories of increasing inclusiveness. Several genera (plural of genus)

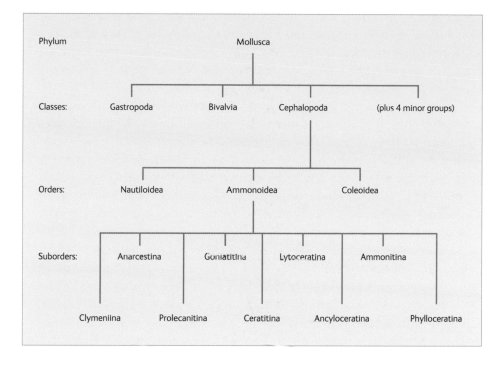

LEFT: Diagram showing the higher-level classification of the ammonites, which belong to one of three orders of cephalopod molluscs.

may be grouped together in a family, several families together in a superfamily and the superfamilies themselves are clustered into classes, two or more of which combine to make the phylum. It all sounds rather complicated, but it does serve a useful function in ordering the almost countless number of species in the most economical way. All the species within any of the units of classification descended from a single, ancestral species. We need not worry about families or superfamilies in this book, but classes often correspond to everyday groups of animals. In the phylum Mollusca – the molluscs, which include the majority of sea animals with shells – the class Gastropoda includes all the snails and the class Bivalvia all the clams (mussels, scallops, etc.). Many of the most familiar fossils, like the trilobites, are classes within larger phyla.

Plants are classified in much the same way as animals, but just to make things difficult the largest units of classification are not generally called phyla but 'divisions'. Many of the plants familiar in the garden – dahlia, chrysanthemum, fuchsia – are generic names which have come into common language. The fossil record of plants is rather patchy compared with that of some animal phyla. Fungi have a most imperfect fossil record, although it is very likely that they have been around for nearly as long as the plants. Recent techniques based on molecular sequencing are providing another way of deriving classifications of both plants and animals; those organisms which have most similar genomes are likely to belong to the same group. Molecular evolutionary 'trees' have solved some long-standing problems in classification for which no relevant fossils have yet been discovered.

In the remainder of this chapter we will examine the commonest and most important kinds of fossils, the kinds that the reader will be able to find when he or she starts a collection. The groups of organisms described have certain special peculiarities that set them apart from all others, and once these features are recognised the group to which the fossil belongs can be confidently identified. Each group is illustrated by a typical species, but of course all the groups contain many species, and any specimen the reader is likely to find will probably differ from the one chosen for illustration in several details. Most accounts of the different kinds of fossils start with the simplest organisms and work towards the most complex, often ending somewhat selfishly with human beings. This can give the impression of a simple evolutionary tree: single cells to many cells, shellfish to fish, and then land vertebrates. This is not correct. Although single-celled organisms are obviously simpler than complex animals like ourselves, many single-celled organisms have continued to evolve actively since they lived in the 'primeval soup'. It is known that some of the unicellular organisms in the sea today are only 1 or 2 million years old (very little in geological terms); in fact they are probably as young as humans themselves. Very simple and very complex organisms have lived side by side for a long time and both have evolved together. True 'living fossils' are really rather rare, and the term can be applied to both simple and highly complex organisms that have outlived the time when the Earth was populated with many more of their kind.

It is easier to describe simple organisms first and move on to complex ones, and we will follow this procedure here, even though this is just a convenience of arrangement. Some of the most important fossils are extremely tiny: these will be discussed in a later chapter, and what follows is concerned with the forms that can be recognised from hand specimens. When they are first chipped out of the rock such fossils will often be partly concealed by the surrounding rock (or matrix). To identify them it is usually essential to clean off most of the enclosing rock, otherwise one can be misled by superficial resemblance. The rule is never to try such cleaning in the field – it nearly always results in a pile of rubble and a frayed temper.

SINGLE-CELLED ANIMALS – PROTISTS

A large and heterogeneous collection of organisms are grouped in the protists, including all those animals that can lead an autonomous existence as a single cell. Most are microscopic, and many have no skeleton and therefore lack a fossil record. The flexible amoeba, which is the protoplasmic 'blob' of popular imagination, is a familiar protist without much potential for fossilisation. A few groups secrete hard tests (shells), which are very common fossils, but again most of these are too small to be easily spotted in the field. They are of great geological importance, however, and we shall return to them in Chapter 10. Such testate protists have a record extending back to the Cambrian. A very few secrete skeletons of (for a single cell) gigantic size, and these also occur in such numbers that they form conspicuous and common fossils. Particularly in rocks of Eocene age the coin-like nummulites are important as rock builders. Nummulites are known from England, but they are most numerous in the Mediterranean and eastwards, where they form limestone formations of great thickness. The pyramids of Egypt are constructed of limestone blocks in which the species *Nummulites gizehensis* is conspicuous. Nummulites are a giant kind of foraminiferan, an important group of rock-forming organisms as far back as the Carboniferous (and with ancestors in Cambrian rocks), and which still form deep-sea oozes today. Most foraminiferans require a microscope for their study. Another group of giant foraminiferans flourished particularly in the Permian period: these are the fusulines. Instead of being about the size and shape of a coin, the fusulines are spindle-shaped, with a round cross-section tapering at both ends; fusuline limestones again form thick rock sequences, particularly in Russia and

LEFT: The giant foraminiferan *Nummulites gizehensis* (nummulite), Eocene, Egypt; 'coins' up to 4 cm (1¹/₂ in) across. Numerous specimens of this species are preserved together, actually forming the rock. The small, lentil-shaped fossils, also visible ,are a separate form of the same species.

the Orient. Both fusulines and nummulites are divided into many small chambers internally, and details of their internal structure are used to classify them. Both groups seem to have flourished in exceptionally warm, shallow seas in the tropics of the Tethys, the sea that once ran through the centre of Europe and eastwards to the Himalayas.

THE SPONGES – PHYLUM PORIFERA

The sponges are an important group of many-celled organisms, with the individual cells specialised for particular functions but rather loosely aggregated. Some sponges have the property of being able to 'rebuild' their colonies of cells if they are mashed up. Marine sponges have a long fossil record from the Cambrian onwards, and at many localities they are abundant enough to be important rock-formers. They have been prominent reef-building organisms too, often in association with bryozoans. Not all sponges contain hard parts capable of being fossilised. The skeletal elements of those with hard parts are fine, often branching elements called spicules. The spicules may be only loosely associated within the sponge tissues, or may be more or less fused to form rigid skeletons. Two major divisions of sponges are distinguished on the composition of the spicules: the glass sponges have a skeleton composed of silica, one of the few examples in nature where this material is used as a basis for a skeleton, while the calcareous sponges use calcium carbonate for their spicules. Both kinds have Cambrian representatives and both flourish today, the glass sponges being especially numerous in deep-sea environments. The sponge skeleton is highly variable in shape: many are cup or flask shaped, others spherical or cauliflower-like, while some form flat plates folded together, or are encrusting. A successful group of sponges has taken to boring into the shells of other organisms. In detail, sponges are distinguished by the form of

RIGHT: The hexactinellid sponge *Coeloptychium agaricoides*, Cretaceous, Westphalia, Germany; 'cap' is 8 cm (3 in) across. These two specimens are beautifully preserved, extracted from a matrix of white chalk, which gives them their colour. Species of *Coeloptychium* can be found in England and elsewhere in Europe.

RIGHT: Glass sponge, *Hydnoceras tuberosum*, Devonian, New York; 20 cm (7$^{1}/_{2}$ in) long. This large sponge is preserved in a fine sandstone. *Hydnoceras* is a Devonian and Carboniferous genus, but glass sponges of a similar general form (but without knobs) have a history going back to the Cambrian, and survive today.

the spicular structure, and the way these combine to form the skeleton, as well as the overall shape. Although sponges are widely distributed in the fossil record they are particularly numerous and easy to collect in Cretaceous rocks in Europe and North America, where dozens of well-preserved species have been described. Sponges are often preserved inside flints, which were originally deposited as hard layers within the Cretaceous chalk. As they are much more resistant to weathering than the enclosing chalk, the flints remain behind when the chalk is eroded away. Flints containing sponges may be incorporated into younger sediments (they are then known as derived fossils), and it is not unusual for such flints to be dug up in suburban gardens around London – a long journey from the Cretaceous seas.

CORALS AND RELATED ANIMALS – PHYLUM CNIDARIA

The Cnidaria (formerly coelenterates) are a varied group of organisms with cells organised into definite tissues, a mouth surrounded by tentacles, and a stomach. The tentacles are arranged in a circlet about the mouth, and the cnidarians are typified by such radial symmetry. The majority of cnidarians have a jelly-like consistency, and the group includes the familiar jellyfish and sea anemones, as well as the smaller hydroids commonly kept in schoolroom freshwater tanks. Cnidarians come in two main designs: they are either attached with the tentacles surrounding the mouth waving like an animal flower in the water (polyp), or free-floating with the tentacles hanging down, often as fine as hair (medusa). Many cnidarians have stinging cells that they use to paralyse their prey. It is surprising to find that the soft-bodied jellyfish have any fossil record at all, but in fact they have the longest one of the phylum. Casts of the stomach cavities have been described from Precambrian rocks in many parts of the world, and many different Precambrian jellyfish have now been named. It is certain that the seas of 600 to 1,000 million years ago swarmed with drifting medusae, just as they do at present. These free-floating cnidarians are therefore both primitive and ancient. Some groups have developed the ability to secrete hard skeletons, of which by far the most important are the corals (Class Anthozoa).

The coral is essentially a sea anemone that supports its body by a skeleton of calcium carbonate. This first happened early in the Ordovician (some believe even earlier), and there were no doubt simple polyps in the Cambrian and Precambrian from which the corals evolved. The secretion of a skeleton gave the corals a great advantage over many other cnidarians; supported thus they could overcome the lack of rigidity of their soft tissues and grow large. Many coral species form branching colonies in which hundreds of individual polyps live on top of their own houses. Other forms remain solitary – a single polyp. From soon after their inception colonial corals started to live together in large masses with one or many species forming large, mound-like, wave-resistant structures – coral reefs. Many organisms other than corals also contribute to the construction of the reef. Fossil reefs are one of the prime sites to look for fossils of many different kinds besides the corals themselves.

Corals can be divided into several major groups. In the Palaeozoic, corals looking generally similar to living reef corals may be only distantly related, if at all, to our present-day fauna. These rugose corals can be solitary, or massive in large reefs; in either case their skeletons are composed of the form of calcium carbonate known as calcite. The finest details of the skeletal structure are well preserved, and the corals can be cut and polished, or

BELOW: Pliocene coral, *Septastraea forbesi* Maryland, USA; 10 cm (4 in) long. A massive coral with irregular knobs, composed of corallites with a diameter less than 0.5 cm (¼ in), with polygonal outlines, and with about 12 prominent septa like the spokes of a wheel. This is a fragment of a much larger colony,

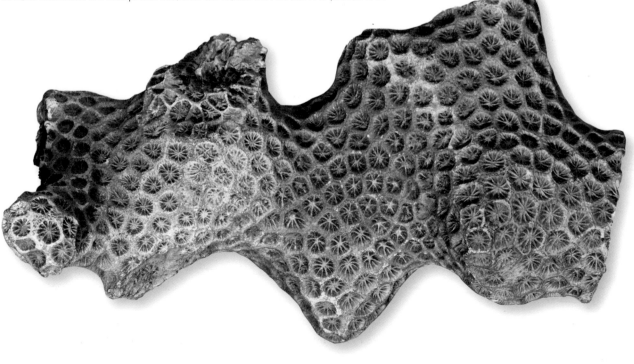

RIGHT: Brain coral, *Coeloria labyrinthiformis* (Scleractinia), Miocene, Antigua; 8 cm (3 in) across. Long meandering corallites with thin septa form a dense meshwork to give this coral an appearance like the labyrinth implied in the name.

ABOVE: Devonian coral, *Heliophyllum corniculum*, length 3 cm (1 in). These silicified specimens have weathered out of the rock. They are small, simple corals with a gently curved form. The deep cup was occupied by the polyp or living coral animal, in life. *Heliophyllum* has numerous species in the Devonian of Europe and North America.

RIGHT: The solitary rugose coral *Cyathophyllum* sp., Devonian, Devon, England; diameter is 6 cm (2½ in). The gaps between the walls of the coral skeleton are filled with calcite, showing up as the lighter colours of the sections. Corals are often best sectioned like this one to reveal the internal structures.

studied in thin sections. The detailed structure of the individual plates that combine together to build the skeleton are the basis for the classification of the corals, as well as the general form. The most obvious of these building elements are plates arranged radially, looking like the spokes of a bicycle wheel in section (septa). In rugose corals the septa usually have a basic four-rayed symmetry in each corallite. Rugose corals are common fossils in rocks of Ordovician to Permian age.

Alongside the rugose corals low cushions or branching masses of a different kind of calcite coral are often found. These have much smaller corallites usually only a millimetre or two across, and the septa typical of rugose corals are absent or inconspicuous. These are the tabulate corals. Both rugose and tabulate corals are extinct.

Corals looking superficially very like the rugose ones are found in Mesozoic and younger rocks, and are important on reefs today (scleractinian corals). Their calcium carbonate skeletons are composed of the mineral aragonite, chemically the same as calcite, but with the atoms arranged in a different way. The septa in these corals are often arranged about a sixfold symmetry so that they are fundamentally different from the Rugosa. Oddly enough, although they are younger than the rugose corals, the aragonite composing them does not preserve very well, and it is easier to find beautifully preserved examples of the older Rugosa. Scleractinian corals are found in small numbers in Triassic rocks, and are abundant in Jurassic, Cretaceous and younger rocks, particularly in past and present tropical regions.

BELOW: Jurassic reef coral, *Stylina alveolata*, Nattheim, Germany; diameter of colony 10 cm (4 in). In life the coral was composed of aragonite, but this recrystallized into calcite – chemically the same as calcite, but with a different crystal structure. Related species of *Stylina* occur in Jurassic limestones in Europe, eastwards into Asia Minor, and in the USA.

THE GRAPTOLITES – PHYLUM HEMICHORDATA

The graptolites are an extinct group of colonial organisms, with a geological record extending from the Cambrian to the Carboniferous. For many years they were regarded as colonial cnidarians, but it is now certain that they are unrelated to the jellyfish and their allies, and in fact are distant cousins to a small group of tube-dwelling organisms with little fossil record, which belong to the minor phylum Hemichordata. Hemichordates are a primitive group that probably share a common ancestor with the chordates (including vertebrates). The graptolites swarmed in the seas of the Ordovician, Silurian and Early Devonian periods. They are usually preserved as flattened impressions, which retain little of their finer detail. The impressions generally show a serrated edge like a saw, each 'tooth' being the crushed tube (theca) that housed an individual of the colony (zooid). The colonies vary widely in shape: some are shrub-like, with numerous slender branches; others have only a few, or even a single branch. The bushy ones (Order Dendroidea – dendroid graptolites) are the more primitive, and were generally rooted to the sea floor. The few-branched forms, which include the Order Graptoloidea, or planktonic graptoloids, were derived from the dendroids at the base of the Ordovician and were successful and prolific until the Early Devonian. They then disappeared, to be survived by their more primitive bushy relatives. The number of branches and the arrangement of the thecae are important in their identification. The graptoloids are one of the important groups used in dating rocks. They evolved rapidly and spread widely, and with a little experience a glance at an assemblage of graptolites on a shale surface can be used to determine the approximate age of deposition of the rock. Most of the later graptolites had only a few branches and, in the Silurian, species with only a single branch tend to dominate the assemblages. One advantage of graptolites is that they occur predominantly in former oceanic environments – the deeper water shales or limestones discussed in Chapter 3. Their preservation in inner shelf habitats is much more unusual – although not unknown. Whether the graptolites were not living near the shore, or whether the conditions were not right for their preservation, are questions that have excited some argument. Whatever the explanation, other organisms, like trilobites or brachiopods, were more numerous in such sites, and have been used to date the rock sequences in the absence of graptolites.

BELOW: Dendroid graptolite. *Rhabdinopora flabelliformis*, Ordovician, North Wales. The large colony of this net graptolite is preserved in a light-coloured shale. Of the many species of *Rhabdinopora* only a few have a colony as regular in shape as this one, which can grow to a length of more than 20 cm (7½ in).

articulated brachiopods rose to prominence only to decline. As with other invertebrates the combination of brachiopod types in an assemblage gives a quick clue to the age. Modern studies rely particularly on the internal structures for their identification: some brachiopods with similar exteriors can have very different 'insides'. But the overall shape of some kinds is distinctive enough for quick recognition – the large spiriferid brachiopod shown on p. 76 is unlikely to be confused with any other, and brachiopods of this type are abundant in the Carboniferous limestone. Above all it is the sheer variety and abundance of brachiopods that give shallow water Palaeozoic assemblages of fossils their distinctive 'feel'.

SEA MATS – PHYLUM BRYOZOA

Bryozoans are an important group of colonial organisms that form encrusting mats on other marine shells or rocks, or branching, leaf-like or hummocky colonies on the scale of a few centimetres. Their colonies can often be seen on seaweeds stranded on the shore, where they resemble a fine, white net attached to the darker weed. They are rather common fossils from the Ordovician onwards, and in spite of their small size can be abundant enough to be important as rock-formers (especially in limestones). Bryozoans found as fossils have calcium carbonate (calcite) skeletons that have a high chance of being preserved, and the fossil forms seem to have been exclusively marine (some living bryozoans live in freshwater). A microscope is essential for their proper study, but even with a hand lens a glance at the surface of a bryozoan colony reveals a number of tiny openings. Each one of these openings was, in life, occupied by an individual of the colony (zooid), a minute animal with tentacles covered in cilia, that entrapped passing micro-organisms and edible particles. Larger colonies would have had hundreds of individual zooids. In more advanced bryozoans of the Mesozoic to Recent individual zooids have been specialised for particular functions – some have become totally modified to curious, snapping structures looking remarkably like miniaturised parrot heads, which may function to prevent the settling of larvae of unwanted alien organisms on the bryozoan colony. The zooids themselves are, of course, not preserved as fossils – we only have their vacated homes. But from a study of detailed thin sections through the colonies we can deduce a good deal about the growth (astogeny) of colonies of bryozoans long since extinct. Although they were filter-feeders like the graptolites, the bryozoans have not taken to free-floating, planktonic existence – they are characteristic benthic organisms. They are divided into a number of groups on the basis of details of structure of the individual 'boxes' that housed the zooids, and the construction of the colony. It is always worth examining the surfaces of fossils like brachiopods or sea urchins to see if the fine matted or delicately branching colonies of

ossil lamp shell,
la maxima, Pliocene,
a, England; adult shells
n) or more long. This
looks today much as it
it had just died. One of
s larger than the other,
alve has a circular
t its apex, through
shy stalk passed
the animal during life.

nestrellina plebia,
oan from the
ous of North Wales.
cimen, the colony
6 cm (2½ in) across.

BRACHIOPODS – PHYLUM BRACHIOPODA

A first collection of marine fossils will almost inevitably include a brachiopod or two. Brachiopods have one of the longest histories and one of the best fossil records of any invertebrate. They are already present in early Cambrian rocks and are still with us today, although living brachiopods tend to be rather inconspicuous in shallow waters. But during the Palaeozoic and Mesozoic they occurred in such profusion in inshore sediments that they are frequently important components of the rocks in which they are found.

Brachiopods have two valves, and this gives them a superficial resemblance to bivalved molluscs. Bivalves and brachiopods really differ in fundamental symmetry, and the differences in the shell reflect even more profound ones in the internal, soft anatomy. Brachiopods have probably always filter-fed, living off small organic particles brought on currents. The particles are harvested by a lophophore covered in cilia – which serves the dual purpose of both creating a current and catching the food. Most brachiopods were attached to the substrate by a stalk, and the hole through which the stalk entered the shell can often be seen on the fossils. This method of food-gathering and the inactive mode of life may seem sufficiently dull for us to anticipate that the brachiopods would have changed little in their long history. It is true that one brachiopod – *Lingula* – is one of the most famous 'living fossils'. *Lingula*-like species, looking much like the living form, can be found in rocks as old as Ordovician. But the other brachiopods have been far from evolutionarily inactive – they have gone through several major proliferations and diversifications, and suffered dramatic major extinctions as well.

RECOGNISI 78

LEFT: Brachiopod, *Spirifer striatus*. Carboniferous, Kildare, Ireland; 5 cm (2 in) across. This brachiopod has a wide hinge line, and the apertural margin is deflected downwards to form a broad 'v'. The radial ribs are more numerous than those of the related brachiopod *Spiriferina*.

RIGHT: The rhynchonellid brachiopod *Cyclothyris difformis*, Cretaceous, Devon, England; 3–3½ cm (1½ in) long. Wide ribbed shells with a small 'beak' projecting from the upper valve, the lower valve is deeply convex. *Cyclothyris* is found in both Europe and North America.

In the Ordovician and Silurian periods they became adapted to life in most marine environments, but were particularly numerous in shallow water habitats, in some cases forming whole banks, as mussels do today. Although generally small (and hence easily collected) some species grew to 10 centimetres (4 inches) long or more. Some brachiopods are smooth, but many became corrugated and ornamented with coarse or fine ribs. The margins of the valves are often wavy, and deeply folded in other species. Long spines on the exterior of the shell were developed, especially during the Carboniferous. Even more profound changes happened in the internal structures of the brachiopods, particularly those concerned with supporting the lophophore – these changed from simple loops to complex 'doubled back' structures, or to fantastic spirals and whorls – all presumably designed to increase the ways of food-gathering, and its efficiency. Quite complicated tooth-and-socket hinges were developed between the valves. Although the variety of brachiopods in the Jurassic and Cretaceous is somewhat less than in the Palaeozoic, they are still very abundant and varied fossils. It has been suggested that their slow decline in the last 100 million years or so has been caused by the commensurate rise in diversity of filter-feeding bivalve molluscs, which ousted them from their former habitats. It is an attractive idea but one difficult to prove; in any case many of the greatest successes of the bivalves have been in life habits that the brachiopods never adopted (burrowing and swimming free, for example).

Brachiopods fall into two major types; the more primitive have shells composed of calcium phosphate plus organic material and hinge development is imperfect. Lingulid brachiopods can be recognised on the rock from their shiny lustre. *Lingulella* is one of the best-known genera. Phosphatic brachiopods were numerous in the Cambrian and Ordovician periods. The other major kind of brachiopods have shells composed of calcium carbonate, and with variously developed hinges, and they comprise a more varied and abundant group of fossils. Through their long history from the Early Cambrian to the present different groups of

ABOVE: F
Terebratu
East Angl
10 cm (4
specimen
did when
its valves
and that
opening a
which a fl
attaching

BELOW: *F*
a net bryz
Carbonife
In this spe
measures

LE
Li
Er
br
ca
lu
m

bryozoans are preserved on their surfaces – they are easily overlooked. Other bryozoan colonies are more immediately conspicuous, particularly the stout, twig-like branches of the Palaeozoic trepostomes, which can make up thick limestone beds, and formed their own 'reefs', or the large, often net-like colonies of the 'fenestrellids' common in the Late Palaeozoic. A relative of the latter, the curious, screw-like *Archimedes* is shown here to the right.

Bryozoans are quite commonly preserved as internal moulds – the calcite skeleton is dissolved away leaving only the sediment fillings of the chambers once occupied by the zooids. In this case they can present a different appearance, a host of little tubes combining together into branching twigs or nets. Such preservation is usual in sandy or silty rocks. The variety of colony structure in the bryozoans exceeds even that in the corals, and ranges from compact humps to cobweb-like branching colonies of exceptional delicacy, and their small size in no way detracts from the beauty and variety of form they display.

ABOVE: Peculiar, screw-like bryozoan, *Archimedes sublaxus*, Carboniferous. Most bryozoa require microscopic examination, but a few form large and distinctive colonies such as this. Numerous *Archimedes* species occur worldwide in rocks of Carboniferous and Permian age. The colonies grow to a length of 5 cm (2 in) though larger species have been described.

MOLLUSCS – PHYLUM MOLLUSCA

The molluscs are one of the most varied, successful and numerous of the invertebrate phyla. Thousands of living species occupy every marine habitat from the shallowest shores to the deepest abyss, and, as every gardener who has tried to protect his or her vegetables against marauding hordes of snails and slugs knows, they have been remarkably successful in making the transition from sea to land. The most primitive molluscs lack a shell, but the most diverse groups, and those that concern us here, have well-developed hard parts that are readily fossilised. Molluscs have the lower part of the body developed as a muscular foot, which may variously function in locomotion, digging or swimming. The molluscs were ultimately derived from a soft-bodied creature somewhat like the living flatworms, but the acquisition of hard parts happened very early on, low in the Cambrian or before, and by the Ordovician the important living classes were well established. A few molluscs, like the slugs and octopuses, have secondarily lost their shells, having developed other means of protection. As with the other phyla, the great span of geological time has seen different groups rise to prominence, then decline, to be replaced by others. But, unlike the brachiopods, the molluscs are probably more diverse today than they have ever been. Measured by their total living weight (biomass) molluscs are one of the most important groups in the whole marine biosphere: the swarms of squid in the oceans are the match of any species of fish. The level of organisation achieved by the most advanced molluscs, for example the octopus, is the most intricate and sophisticated of any invertebrate: one cannot overestimate the importance of the Mollusca in shaping the patterns of marine communities we see today. Different molluscs fill most of the possible ecological roles available to marine organisms: some are voracious hunters, others graze on algae or feed directly on organic material in muddy sediments, others again are filter-feeders. The great majority of molluscs have a minute, planktonic larval stage, a small ciliated object bearing no resemblance to the adult, which drifts as part of the plankton until ready to settle and assume its mature form. Thus even bottom-dwelling molluscs can be dispersed widely over long distances, and are quick to colonise vacant sites that appear in the ocean (new volcanic islands like Surtsey, for example). Most molluscs are small, a few centimetres long, and some are really tiny, but a few species have attained considerable dimensions. The most famous (or notorious) is the giant squid at 15 metres (50 feet) or more in length, but some of the extinct ammonites and nautiloids were of similar dimensions, and were the largest shelled animals ever to have lived.

Because the molluscs are such a large and varied group, with such an extensive fossil record, we shall consider them below in their various classes. Accurate identification of molluscs is a skilled business, and there is a vast literature describing fossil species, but it is easy to recognise the most important types preserved as fossils. The fossil record, particularly in the Cambrian, is still turning up new and fascinating kinds of molluscs, some of them really weird, and a few of these will be touched upon after the most important and familiar groups have been described.

The clams – Class Bivalvia

In clams the body is enclosed in a pair of valves, which in most species are mirror images of one another. The valves are composed of calcium carbonate, and are quite strong in most species, and hence easily fossilised. Between the valves a springy ligament keeps the valves in a gaping attitude – the position used for feeding. If danger threatens, powerful muscles can snap the valves tightly shut, and once a bivalve has closed itself in this way it can be very difficult to force it open – it has 'clammed up'. The strong muscles leave scars on the interior of the shell at the places to which they were attached (usually two in each valve), and these muscle scars may also be seen on fossil shells. To make the hinge strong and efficient there are usually complex arrangements of teeth and sockets there, and the arrangement of these articulating devices is very important in identifying the different kinds of clams. So it is important to find the internal structures in fossil bivalves, as in the brachiopods. The clams

RIGHT: Trigoniid bivalve. *Scabrotrigonia thoracica*, Cretaceous, Tennessee, USA; 5.5 cm (2 in) across. This is exceptionally well-preserved showing the finest details of the shell structure, and the radial ribs 'chopped up' into little knobs are characteristic of the trigoniid bivalves. Inside there are very few, powerful hinge teeth.

LEFT: Carboniferous bivalve, *Aviculopecten planoradiatus*, Derbyshire, England; 3 cm (1 in) long. The original colour banding is preserved, showing as broad, V-shaped patterns. The single valves are preserved in a fine-grained limestone. Pectinid shells have grooves widening towards the margin, and, at the apex, flattened 'ears'.

use their foot for digging and movement generally. The group has adapted to a range of marine and freshwater habitats where they are often filter-feeders. Their fossils often occur gregariously as they lived, forming beds largely composed of fossil shells. Many living species burrow into sand or mud, sometimes to a considerable depth – these species maintain contact with the sea by means of long siphons, tubes that permit the passage in and out of water and bring to the animal both the necessary oxygen and the small organic particles on which it feeds. Other species are attached to rocks by means of tough threads (byssus) that enable them to hang on even in turbulent situations; mussels can colonise the most inhospitable rock surfaces in this way. A few bivalves have become free-swimming – such pectinids can escape predators rapidly by 'clapping' their valves together. Still others have taken to burrowing into wood or even into limestone, and fossils of these curious animals can be found lying in their home-made burrows (see p. 23). With such a wide range of adaptations it is not surprising to find that the shapes of bivalves are highly varied – some are globular, others flat and plate-like, some like the razor shell (*Ensis*) have become greatly elongated to aid burrowing, and in some forms the usual 'mirror image' symmetry has been lost. The thicker-shelled species often carry a distinctive sculpture, which is also important in identifying fossil species.

ABOVE: Tellin bivalve, *Tellinella rostralis*. Eocene, Belgium; 3 cm (1 in) long. This species has an elongate form extended at one end into a shovel-like tongue; fine concentric ribs are nearly parallel to its margin. Species resembling this one are numerous in Tertiary marine formations, and similar species live today in sandy sea bottoms.

Most bivalve fossils are a few centimetres long; the ideal size for collecting. But a few giants are known. The Cretaceous genus *Inoceramus* sometimes grew to well over a metre in length; fragments of this particular genus are frequently important components of the soft, white Cretaceous limestone known as chalk. The giant clam (*Tridacna*) of modern reefs is a familiar living goliath, but reports of it greedily trapping divers are legends of dubious veracity.

The bivalves have a long geological history, with a few doubtful species known even in Cambrian strata, but like many other molluscan groups they really become established and diverse during the Ordovician. By the end of the Ordovician they had already radiated into many of the niches they occupy today. Their story from then on is one of slow, but steady, increase and diversification. Bivalves seem generally to have evolved at a slow canter rather than a brisk gallop (see ammonites p. 88) and some living bivalves have a very long ancestry; the small-sized genus *Nucula* has relatives in Ordovician rocks not very different from Recent species. But the ancestors of many of the modern bivalves arose during the Mesozoic, and unlike many of the animals in this book the bivalves were affected in a much less dramatic way by the Mesozoic–Tertiary extinction events that extirpated so many other major elements in the marine fauna. In spite of their general conservatism the bivalves did produce some short-lived, bizarre forms with no living survivors. Most extraordinary of these are the Cretaceous rudists (see p. 55), a group in which one valve became modified to a long cone, on which the other valve rested like a lid, the whole effect being most un-clammish.

Some clams have shells composed of the form of calcium carbonate known as aragonite (like scleractinian corals) which is easily dissolved away when they are entombed in sediment: this leaves the clam preserved as an impression on the sediment of its internal and external surfaces (internal and external moulds). It is important to collect both 'halves' of the fossil to get an accurate representation of the shell – the external mould will preserve the overall shape and sculptural details, while the internal mould shows what the teeth and muscle scars were like.

The reader may be confused to find what we here call bivalves, referred to in other books as pelecypods or lamellibranchs – these are just different names for the same animals. The name accepted by most scientists today for the two-shelled class of molluscs is also fortunately the simplest – Bivalvia.

The snails – Class Gastropoda

'Slugs and snails and puppy dogs' tails' – there seems to be something in the popular imagination that finds snails slightly repulsive. But this class of molluscs includes not only the greatest number of living molluscan species, including those that have most successfully colonised land, but also some of the most beautiful examples of natural engineering in the zoological world. Some of their shells have a financial value that may even be out of proportion to their aesthetic qualities. The gastropods are molluscs with a single, usually helically coiled shell, with the foot modified into an efficient creeping organ, with a head, usually with eyes and tentacles, and with a rasp-like feeding organ (radula) composed of a series of pointed teeth.

FAR LEFT: The gastropod *Desmoulia conglobata*, Pliocene, Italy; 4 cm (1½ in) long. As befits its relatively young geological age, the specimen is preserved with its original shell material. It has been sliced longways to show the internal structure. Note the *columella*, which is twisted like a corkscrew.

LEFT: Eocene gastropod, *Voluta muricina*, Epernay, France; 7 cm (2½ in) long. The species is beautifully preserved, retaining something of its original lustre, and all the fine details of its ornament. It is distinguished by its tall spire, elongate aperture, prominent spines, but without the spiral ridges seen on many species.

LEFT: Gastropod *Poleumita discors*. Silurian, Dudley, England; 6 cm (2½ in) across. This is a very ancient snail with a flat, nearly disc-like form, with fine growth lines and occasional coarser knobs. The topmost whorls of the shell are broken off. Gastropods of this kind are typical of Ordovician and Silurian strata.

ABOVE LEFT: Gastropod, *Harpagodes wrightii*, Jurassic, Gloucestershire, England; 15 cm (6 in) across. This large gastropod is preserved in an oolitic limestone. The long, stout spines are an unusual feature distinguishing this species from other gastropods. This species is a real rarity.

ABOVE CENTRE: Sundial shell, *Architectonica millegranosa*, Pliocene, Orciano, Italy; 3 cm (1 in) across. The low spire of this species, with a minutely beaded ornament, and the sharp rib around its outside edge, are features that discriminate this species from other gastropods. This species is one of a fairly large genus that survives today.

ABOVE RIGHT: Limpet gastropod, *Symmetrocapulus tessonii*, Jurassic, Les Moutiers en Cinglais, France; 11 cm (4 in) long. The fossil shows the original concentric growth wrinkles well. Fossil limpets of this kind are generally rare as fossils, and this specimen is very rare indeed, and large for a limpet-like mollusc.

The gastropods have become adapted to a very wide range of habitats, from high mountain streams to deep oceans, and each habitat type has its own species confined to it. Freshwater gastropods are different from marine ones, and within the marine habitat itself the gastropods are strongly zoned ecologically, so that even on the same shore different species will be found in different areas according to their relation to the tide marks, degree of exposure, their diet, and so on. Usually such fine habitat details are not preserved in the fossil record, and the gastropods we find are a jumble of shells displaced from their original microhabitats. Many gastropods are grazers: they use their radula to rasp away at algae. Some feed more or less directly on sediment, from which they extract edible particles. There are a large number of living species that are predators, some species using the modified radula as a kind of poison dart, others employing it to bore neat, perfectly round holes through the shells of prey (often bivalves) in order to get at the nutritious interior. Some of these predators are specialists concentrating on one particular prey type. It will now be clear why you can sometimes find so many different fossil gastropod shells together in a single fossil deposit.

The gastropod shell can vary from a millimetre in length to several tens of centimetres; many are thick and robust, others delicate and fragile. While the majority are coiled in an upward spiral, some are modified into simple, cap-like forms (like the limpets), and others are coiled in a flat plane, like a ram's horn. Some genera have high spires, with many turns of the shell visible externally; others have low, broad spires in which the last whorl overlaps the earlier ones. Perhaps more than any other group of molluscs the gastropods are remarkable for the variety and beauty of the external sculpture on the shell, which may be covered with a delicate tracery of ribs and lines, or stout spines, or fine prickles. Many species have the aperture flared, or extended into a long tube (siphonate forms). Sometimes you can find a lid (operculum) preserved as a fossil; this closed the aperture when the mollusc withdrew into its shell. All these characters are used in the identification of fossil species, but of course the colour patterns that may be characteristic of living species are not available in the great majority of fossils. Like some of the bivalves, some gastropods had shells composed of the mineral aragonite, and these are usually preserved as moulds; a cast taken from the external mould will replicate the fine detail of the external sculpture.

There are small, helically coiled shells present even in Cambrian rocks, and it seems

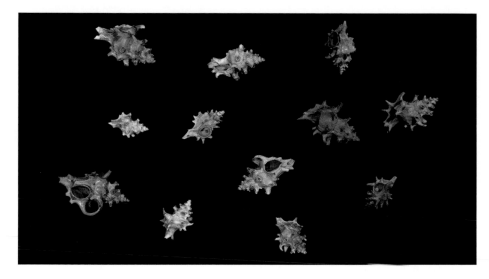

LEFT: *Typhis pungens*, a gastropod, Eocene; largest is 3 cm (1 in) long. The *Typhis* shown here demonstrate the different sizes gastropods that one species can attain. There are a number of *Typhis* species, with various numbers and arrangements of the spines. *Typhis* still lives today, e.g. in the seas around Japan.

BELOW LEFT: Pliocene gastropod, *Busycon contrarium*, Florida, USA; 6 cm (2½ in) long. This gastropod is unusual as its aperture lies on its left-hand side, not on the right as in most other species. Its low spire and long, projecting apertural extension are also characteristic. The species still lives in the seas off Florida today.

BELOW RIGHT: Eocene gastropod *Athleta spinosa*, Grignon, France; 2-4 cm (1–1½ in) long. Gastropods of this type are characterized by strong ribs produced into spines, a long aperture, and relatively low spire. There are many similar species, widespread in Tertiary times.

likely that the gastropods diverged from the other molluscs late in the Precambrian. By the Ordovician the gastropods were a varied group present in a variety of shallow water habitats. As might be expected, most of the Palaeozoic gastropods belong to primitive groups: a few of these primitive snails survive today as inconspicuous members of the Recent fauna. By the Carboniferous many of the shapes we see in living gastropods can be matched in the fossils, but despite these similarities the majority of the Palaeozoic forms were not closely related to their living analogues; this is another example of similar-looking forms evolving independently probably in response to similar life habits. During the Mesozoic, forms ancestral to many of the living gastropods evolved. In particular, the adoption of new predatory habits in the Cretaceous was a departure for the gastropods, and was in part responsible for a proliferation of the group unmatched in their Palaeozoic history and may have stimulated the 'Mesozoic marine revolution'. They continued to evolve with undiminished vigour through the Tertiary (Cenozoic), and their fossil remains are nowhere more abundant than in the 'crags' of later Tertiary age. The gastropods are one of many groups that record the faunal changes connected with the advance and retreat of the ice-sheets during the Pleistocene. The invasion of the non-marine habitat probably first happened in the Carboniferous, but relatives of the living land snails are rare before the Cretaceous, at which time the familiar *Helix* made its first appearance. The slugs were derived from this group by reduction of the shell, at which stage they become distinctly less qualified to have a fossil record.

Gastropods can leave other evidence of their activities. Grazing species leave characteristic winding trails, and these have been tentatively identified as fossils, but with the usual caution that the trail makers themselves are never apparently preserved at the end of their tracks.

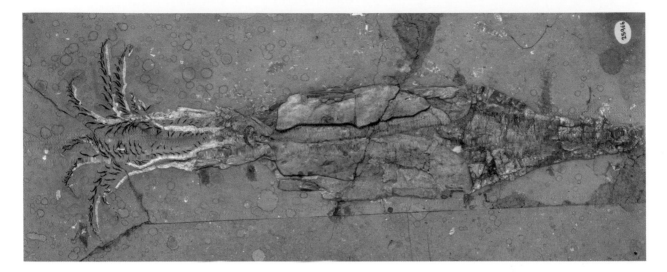

ABOVE: Fossil squid *Belemnotheutis antiquus*, from Christian Malford, Wiltshire, England. Both the body and tentacles are intact in this 160 million year old specimen, which is very rare as the soft parts usually decay very quickly.

Nautiloids, ammonoids, belemnoids, squids – Class Cephalopoda

The cephalopods include the most complex and 'advanced' of the molluscs, and are a group of the greatest geological, as well as biological interest. They have muscular tentacles in a well-developed head region, highly efficient eyes, which are similar in construction to those of vertebrates (although obviously independently derived), and they feed with the aid of strong, beak-like jaws. They are predators, and probably always have been, and it may have been the adoption of hunting habits that favoured the development of high intelligence. Certainly, living cephalopods have a sophisticated nervous system and a relatively large 'brain': octopuses seem to be capable of very rapid learning. Perhaps H.G. Wells made the right choice when he cast octopus-like animals in the role of intelligent alien invaders in *The War of the Worlds*.

The cephalopods evolved from another (possible gastropod-like) mollusc at some time during the Cambrian, but their earliest history is little known. Like most other molluscan groups they rapidly diversified in the Ordovician. These early nautiloids were probably predators also, and if this were so they may have been among the first rapidly moving, efficient hunters in the sea. Their impact on the marine communities of the Ordovician must have been profound. The presence of efficient predators would have acted as a stimulus in the evolution of other groups: even algal grazers would be compelled to evolve protective devices or rapid reproductive strategies to outpace predatory depredations. We have seen how many animal groups diversified during the Ordovician; the rapid evolution of nautiloid cephalopods at the same time may be more than coincidence.

Rapid movement in cephalopods is achieved by expulsion of water from a muscular funnel beneath the head. Most of the living cephalopods also have an ink sac that injects a smoky fluid into the water when the animal is threatened, under cover of which they can make their own jet-propelled escape. Since living *Nautilus* does not have an ink sac, this must have been a protective device evolved at a later stage in cephalopod evolution. Many cephalopod species behave gregariously today, travelling in swarms in pursuit of their favourite prey (often small crustaceans). Fossil cephalopods belonging to one species are often found together in large numbers and this may reflect similar gregarious habits, but there are other possible explanations – for example, concentrations of fossil shells may have been sorted by currents.

NAUTILOIDS

The earliest cephalopods found as fossils are nautiloids and they also have the longest history, because the living pearly *Nautilus* belongs to the same group. The single Recent genus scarcely reflects the diversity attained by the group in Palaeozoic seas, and more than a hundred different genera have been described from the Early Palaeozoic alone. Sections through the living *Nautilus* (see below) are often sold as ornaments, and well illustrate the distinctive features of nautiloid hard parts. The body of the animal occupied the cavity at the end of the spiral – the living chamber. Behind the living chamber the rest of the shell is divided into smaller chambers. As the animal grew it secreted a wall (septum) between the body chamber and the chamber immediately behind – walled off part of its home as it were. Running along the middle part of the shell there is a narrow tube – the siphuncle – that connects the living chamber with the earlier parts of the shell. The empty chambers are usually supposed to have been filled with gas, which help to give the animal buoyancy, and via the siphuncle the animal can vary its buoyancy to control its position in the water column. Some of the early nautiloids deposit calcium carbonate in the voided chambers, which may also have been connected with controlling buoyancy. Where the septum meets the body wall it does so in a smooth curve. Internal moulds of fossil nautiloids often reveal a series of such lines marking the boundaries of the chambers – these are known as suture lines. All nautiloids have simple suture lines.

BELOW LEFT: A living nautilus, *Nautilus pompilius*. The chambered nautilus occurs at depths to several hundred metres in some area of the Indo-West Pacific. It is a living fossil, surviving virtually unchanged for hundreds of millions of years.

BELOW RIGHT: The living pearly *Nautilus*; a section showing its internal division into chambers.

When they first appear in the Early Ordovician the majority of nautiloid shells are straight or slightly curved; they are 'unwound' forms. Some of these straight orthocone nautiloids achieve considerable dimensions, several metres long, and they must have been formidable predators on the other marine animals of the time. Quite early in their history some partly coiled or even tightly coiled species evolved, and the various coiling types seem to have coexisted successfully side by side. Some of the early nautiloids occurred in such abundance that they are conspicuous enough to form an appreciable part of limestone formations – the 'Orthoceras Limestone' (Ordovician) is one of these, widely distributed through Scandinavia. The nautiloids achieved their widest range of adaptations and greatest variety of form in the Ordovician and Silurian periods, with various coiled forms, straight, pipe-like shapes, and curious dumpy species with restricted apertures that may have adopted a sluggish (possibly filter-feeling) mode of life. Thin sections show a great variety of internal structures important in accurate identification. In the Devonian period the nautiloids were still abundant and varied, but they suffered a slow eclipse coincident with the rise of the ammonoids. Nevertheless, unlike many of their Palaeozoic companions, they survived the Late Permian extinction, and the ancestors of the living *Nautilus* even underwent a minor evolutionary burst in the Mesozoic, where forms quite similar to the pearly *Nautilus* can be common fossils. And in the end the nautiloids even survived the ammonites, the molluscan group that evolved more rapidly and more spectacularly than any other.

AMMONOIDS ('AMMONITES')

The ammonoids were derived from the nautiloids probably during the Early Devonian, and from the Carboniferous until the Cretaceous are among the most abundant of fossil groups, in some rock types dominating to the exclusion of most other members of the fauna. Ammonoids differ from nautiloids in the suture lines being wavy or crimped; this of course reflects an elaboration of the outer part of the walls (septa) separating the chambers in the earlier part of the shell. The siphuncle in most ammonoids runs not through the middle of the whorl but along the outer edge. There has been a lot of debate about the reasons for producing complicated patterns on the suture line; what advantage would this have given the ammonoids over their nautiloid ancestors that allowed for the explosive bursts of ammonoid evolution? It cannot be coincidence that the folding of the septal walls occurs at the point where they meet the body shell of the animal; this is a point of relative weakness, and all good joinery benefits from strengthening the joint. The earlier ammonoids had gently wavy sutures, and many of the Mesozoic ones had sutures almost incredibly folded and contorted, so it looks as if natural selection were generally favouring increase in the elaboration of the folds. One simple explanation, which is attractive and plausible, is that the strengthening enabled the ammonoid shell to withstand the hydrostatic pressure at relatively great depths in the ocean – they need not be confined to the surface waters or to relatively shallow depths. Certainly individual ammonoid species became extremely widespread, and oceans were not a barrier to their distribution. They became masters of the pelagic realm, possibly swimming in schools like their distant relatives the squids. Like living pelagic animals they had preferences for particular water temperatures; different types of ammonoids were found in high or low latitudes. Since the disposition of continents has changed, these 'faunal realms' offer a method of deducing the distribution of past climatic belts.

LEFT: Jurassic nautiloid, *Cenoceras pseudolineatus*, Dorset, England; 7 cm (2½ in) across. Specimen largely preserved as an internal mould. A section cut and polished through the specimen (ABOVE) shows the internal chambers filled or partly filled with calcite. The internal mould shows the gently-curving suture lines.

RIGHT: The heteromorph ammonite, *Scaphites nodosus*, Cretaceous, Badlands, South Dakota, USA, 8 cm (3 in) across. This specimen has retained much of its original shell, giving it a beautiful pearly lustre. The species is one of the later ammonites, in which the normal ammonite plane spiral has begun to 'unwind' in various ways.

BELOW LEFT: Ceratite ammonoid, *Ceratites nodosus*, Triassic, Göttingen, Germany; 6 cm (2½ in) across. This is an internal mould in limestone, and shows the suture lines – the boundaries between chambers on the inside of the shell. Such ammonites are plentiful in marine limestones of the right age across continental Europe, and related forms occur in America.

BELOW RIGHT: Carboniferous goniatite, *Goniatites crenistria*, Derbyshire, England; 5 cm (2 in) across. This handsome goniatite is an internal mould preserved in limestone, revealing the suture lines, which have a distinctive zigzag form. The aperture is broken off along one such suture, giving the specimen a pointed margin.

Whatever the reasons for their change from their nautiloid ancestors, the ammonoids were an enormously successful group: thousands of different species have been described, and their variety is so bewildering that many specialists have devoted their lives to studying only the ammonoids of a particular, short time period. The changes that they underwent are an infinite set of variations on a relatively limited number of themes. Most important, perhaps, are the characteristics of the suture lines, which are displayed on internal moulds. The mature shell size also varies from small species a few centimetres across to giants of a metre or more in diameter. Some species grew in such a way that the last whorl overlaps the inner ones; often these species become flat and discus shaped, and it has been suggested that this was a more 'streamlined' shape for active swimmers. Other species have squat whorls, the whole ammonoid being so tightly rolled up as to be almost spherical. The exterior surface of most ammonoids is covered with ribbing – dense on some species, sparse on others – the ribs often split into two or more smaller ribs as they pass over the back of the whorls. Many ammonoids additionally carry spines, warts, tubercles or lumpy excrescences, so that the large shells can look positively burdened with sculpture. It is difficult to imagine that such species were rapid swimmers. Others, particularly the discus-shaped forms, were smooth externally. All these features are used in the classification of the group. A particularly puzzling aspect of the ammonoids is a great resemblance between external shell

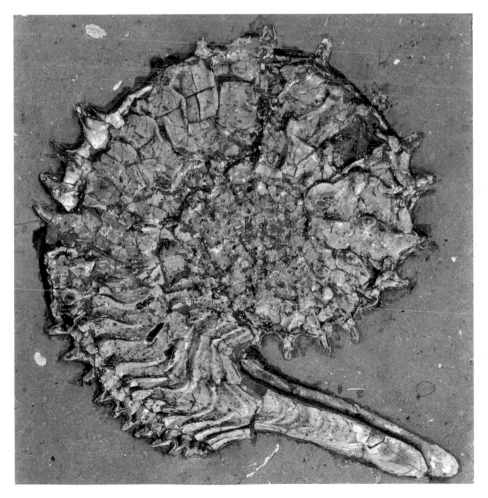

LEFT: Ammonite, *Kosmoceras acutistriatum*, Jurassic, Wiltshire, England; 9 cm (3½ in) across. This ammonite occurs in a fine-grained grey shale, but has been severely flattened. The original lustre of the shell has been retained. This species is particularly distinctive because of the extended flanges at its apertural margin (lappets).

features that can be produced at different times by otherwise unrelated ammonoids. Such homoeomorphs are presumably produced in response to very similar life habits. The origins of a particular homoeomorph can usually be deduced by studying the internal features (in particular the suture lines), or by tracing its derivation from geologically older species in underlying beds.

It is only in relatively recent years that two sexes have been recognised in ammonoid species. They were originally thought to have been two closely related species occurring together in the same rocks, but these 'pairs' were so consistently found together that it became more and more probable that they were sexual forms of the same species. The smaller of the two is considered to be the male, which also tends to have rather stronger ribbing, and sometimes a differently shaped aperture from the larger female.

The early ammonoids of the Devonian (*Anarcestes* and allied genera) have only gently sinuous suture lines. In the Carboniferous a great variety of forms with highly zigzag suture lines are a distinctive group of ammonoids usually known as goniatites. By the Permian–Triassic the suture lines of many ammonites had begun to assume the highly crimped and complex form that was to characterise much of their later history; the ceratites of the Triassic combined broad loops and tight folds in a distinctive pattern (see p. 90). It was in the Jurassic that the ammonoids achieved their greatest flowering, when the clays and shales of the period may be solid with the remains of their shells. The whole gamut of shell shapes, ornaments and sutural complexity was present during this time. Their rise was not, however, one of simple and progressive increase in variety. At several times, notably at the end of

RIGHT: Jurassic ammonite, *Promicroceras planicosta*, Lias, Somerset, England; 2 cm (1 in) across. Numerous small specimens are preserved higgledy-piggledy in an impure limestone. Some retain their pinkish shell, others show the dark internal moulds. Small size, unbranched and strong ribs, giving the shell a ramshorn appearance, are important characters.

the Palaeozoic, they suffered massive and largely unexplained extinctions, a few survivors giving rise to the variety of forms that followed. In spite of their success they seemed to be vulnerable to extirpation in a way which some of the less imaginative molluscs, like the bivalves, seemed to be immune. The group persisted, successfully, into the Cretaceous, at the end of which period the whole group, apparently rather suddenly, became extinct. We shall return to this sudden death of an important group in Chapter 7.

No description of the ammonoids is complete without mentioning the heteromorphs. These are forms that abandoned the usual plane spiral mode of coiling, and instead became partially or even completely uncoiled, or became twisted in some other fashion. Some forms (*Turrilites*) adopted the helical spire, and were it not for the obligatory suture lines it might be possible to mistake these species for large gastropods. In others the coils became loose, like a watch spring. Still others developed a bizarre backward hook in their mature stage. The extreme heteromorphs are perhaps to be found in the genus *Nipponites*, which looks like a tangle of whorls where any obvious semblance of symmetrical coiling seems to have been lost, and *Baculites*, which is virtually straight after its earliest whorls. Most of these heteromorphs were derived from 'normal' ammonoids, but there is one famous example where an uncoiled form actually gave rise to a conventional-looking ammonoid by coiling up again! Heteromorphs are particularly common in, but by no means confined to, Cretaceous rocks. At one time it was supposed that the ammonoids were suffering from 'racial senescence' at that time and that the uncoiling represented a kind of genetic exhaustion. This resulted in their reversion to resemble some of their earliest ancestors (the straight nautiloids) and squared with the observation that some of the heteromorphs even reverted to simple, wavy sutures again. However, the heteromorphs were extremely successful and widespread, and they are accompanied by, and even survived by, other species with a perfectly usual appearance.

BELOW: The Cretaceous genus *Nipponites*. About x1½. The acme of 'unrolling' in the ammonites.

The explanation lies in something less mysterious than 'racial senescence'. Far from being an expression of decline the heteromorphs show the ammonoids adopting new (and successful) life habits. Some of them may have become bottom-living, crawling hunters for which a gastropod-like shell would have been more appropriate. Loss of active swimming habits may have rendered the complex suture lines superfluous. But their displacement from a dominant role in the pelagic habitat may have accompanied the rise of squid, sepioids, and octopods.

Ammonoids, heteromorphs included, are almost without parallel as stratigraphic indicators. They evolved rapidly and spread widely, and have a range of distinctive characters to help investigators in their identifications. Study of evolving populations of ammonoids has produced very fine subdivisions of Jurassic and Cretaceous rocks, and their only disadvantage, curiously enough, is the feature that makes them so attractive to collectors. They are large, and hence the chances of recovering complete specimens from all boreholes is relatively low. Microfossils (Chapter 10) are often used in their stead for subsurface work.

ABOVE: Belemnites, *Acrocoelites subtenuis*, Jurassic, Yorkshire, England; longest specimen 9 cm (3½ in) long. This fine group of belemnites is preserved on a soft shale. The belemnite guards are composed of calcite. There are very many belemnite species in Jurassic and Cretaceous rocks, sometimes occurring in vast numbers.

BELEMNOIDS

The cigar-shaped belemnoids ('belemnites') are common fossils accompanying the ammonoids in Jurassic and Cretaceous rocks. The fossil is an internal skeleton (enclosed within the body) of a squid-like animal, in a way comparable with the cuttlebone of the living cuttlefish. The solid calcite of this internal skeleton makes the belemnoid a resistant fossil, and fragments are common survivors of erosion, often picked up on 'Recent' beaches. The chambered shell (phragmocone) is less conspicuous – it is tucked into the broad end of the fossil guard, where the series of closely spaced septa reveal the cephalopod nature of these otherwise somewhat featureless fossils. Belemnites vary from small fossils a centimetre or two long to large specimens tens of centimetres long: of course these are only a fraction of the size of the living animal, with their tentacles extending well beyond the guard. The belemnoids disappear from the fossil record at the end of the Cretaceous, but some of the group gave rise to living squid-like animals, and so they are not to be regarded as extinct in the same, final way as the ammonoids.

Other molluscan classes

Some molluscs are found as fossils that belong to groups other than the familiar ones described previously. Although they are rarer fossils, some of them are of such interest that they are worth mentioning here. One of the most exciting groups is the Monoplacophora. These are a group of extremely primitive molluscs, which are found in any abundance only in Lower Palaeozoic rocks. They are usually simple cap-shaped shells looking somewhat like limpets, but on their internal surfaces they carry a series of paired muscle impressions. Many authorities regard the Monoplacophora as lying at the root of the other molluscan groups; gastropods, cephalopods and even bivalves may have been derived from them. For many years they were known only from fossil representatives and were believed to be extinct. It was an amazing discovery to find that monoplacophorans were still alive and well – living ones were dredged from the deep ocean in the early 1950s. Not only that, the living form had hardly changed from its Silurian predecessor (see p. 176).

The monoplacophorans really reached their acme in the Cambrian, with curious forms having shapes that have never been paralleled in other molluscs.

Another odd group that did not survive the Palaeozoic was the rostroconchs. These look rather like bivalves without a hinge. They really are extinct, but managed to compete successfully with the bivalves for a considerable time. Finally, the tusk shells (Scaphopoda) are a long-lived group with a record stretching from the Ordovician to the present. They are the only molluscs with a truly tubular shell, usually gently curved, with an opening at both ends. They live today with the broad end buried in the sediment, where they forage for food using small prehensile filaments. Their habits have probably always been similar, and if survival is to be taken as a measure of success, their conservative way of life has ensured them a leading place in the evolutionary marathon. Other animals described in this book have occupied a specific ecological niche where they can quietly pursue their own speciality without embarking on any spectacular radiation in the manner of the ammonoids. It is a curious paradox that often the most spectacularly successful and numerous organisms are also those with a finite geological record.

THE ECHINODERMS – PHYLUM ECHINODERMATA

Sea urchins, starfish, sea lilies and sea cucumbers are living echinoderms – the 'spiny skinned' animals, according to the Greek name of the phylum. It is an appropriate name, for most echinoderms do feel prickly to the touch, and the sea urchins are equipped with fearsome spines. The skeleton of all echinoderms except the sea cucumbers is a relatively strong assembly of calcite plates; the animals are built from an interlocking mosaic of such plates, and mostly the skeletons are rigid enough to have a high chance of fossilisation. The geological record of the phylum is accordingly excellent. The echinoderm skeleton is not a truly external one, like that of the molluscs, for it is surrounded on the outside by a thin skin of living tissue. The mosaic of calcite plates is a particular feature of the phylum, and serves as one of the characters linking together such dissimilar-looking animals as starfish and sea lilies. Another unique feature is the water vascular system, a system of internal plumbing that drives the tube feet: mobile, club-shaped sacs arranged in serried ranks on the outside of the animal, which operate in harmony to convey food to the mouth, or in locomotion. They have a well-developed nervous system. Most echinoderms also have a unique five-rayed radial symmetry in the overall shape of the skeleton. This fivefold (pentameral) symmetry seems a peculiar number to be present over such a wide range of organisms. Why not four, six or thirteen? The answer is still not clear, but what is certain is that the five-rayed plan was established even in the Cambrian in a number of diverse echinoderms; it is evidently highly functional.

In detail the echinoderms show evidence of bilateral (mirror image) symmetry, from which fivefold symmetry was probably derived. Once acquired it is rarely lost, even in echinoderms like sea urchins that (perhaps) could also function on a different symmetry pattern. Each plate of an echinoderm is composed of a single crystal of calcite, whereas all of the organisms we have discussed above have skeletons composed of felt-like masses of tiny crystals. Broken echinoderm plates 'catch the light' – the single crystals break along cleavage planes and so present a uniform reflective surface. Many rocks are composed of a high proportion of echinoderm debris, which is easily recognised because of this optical property. Through their long history the echinoderms have never left the sea, but within their preferred medium they have adopted most of the life habits available to marine organisms. They include grazers feeding on simple plants, scavengers, mud eaters, filter-feeders extracting micro-organisms from currents, and some efficient and versatile hunters. Like other marine organisms they have a planktonic larval stage that assists their dispersal, but almost all adult echinoderms lead a bottom-dwelling existence; a few sea lilies have successfully cast off their anchorage to the sea floor and become widespread swimmers.

The echinoderms are such a varied and important group that the different kinds have to be considered separately. I should add that some of the most interesting echinoderms are also some of the rarest – peculiar, plated animals that do not fit comfortably into the array of living forms. Many of these occur in the Cambrian – it is as if the echinoderms tried out various designs before settling for the successful models that mostly survive today. The echinoderms are a phylum closely allied to the Chordates – including the vertebrates – and both phyla probably had a common ancestor in the Precambrian.

Sea lilies – Class Crinoidea, and other stemmed echinoderms

Crinoids are abundant and important fossils from the Ordovician to the Tertiary. They are still abundant at present, although a little less common in shallow water sites than they were in the Palaeozoic and Mesozoic, but they are conspicuous and varied components of deep water faunas. The great majority of the group have long stalks, which are anchored to the bottom. The main part of the animal consists of a cup (or calyx) to which the stalk is attached at its upper end, and from the top of the calyx stretch long arms, which are five in number or more usually a multiple of five. The arms are often repeatedly branched. The distal parts of the arms carry fine pinnules, which are instrumental in gathering the fine, water-borne food such as micro-organisms on which the animal feeds. Since crinoids often occur together in large numbers ('gardens') with their arms waving in the currents, it is easy to see how they came to acquire their botanical analogy. Food grooves in the arms channel the food to the mouth, which lies in the centre of the calyx. Stem, calyx, and arms are all made of calcite plates. The stems are often common fossils, even when the calyx cannot be found – many of the 'crinoidal limestones' are composed largely of stem debris. The individual plates of the stem slot together like a stack of coins; they are termed ossicles. These can be round, or five-sided, up to a centimetre or so across, and have a small hole in their centre; the stems can look like corals to the casual observer, but where they are broken they show the typical echinoderm reflective surfaces. Detailed classification of the crinoids is based principally on the way the calcite plates are arranged on the calyx with the number and branching patterns of the arms, the external sculpture on the calyx, and important internal features of the cup connected with the nervous and respiratory systems. In detail they are a very complicated group.

Complete specimens of fossil crinoids are rather rare, and deserve pride of place in anyone's collection. Complete cups are more common – many of them were evidently rigid, and they are easily preserved as fossils. There are famous examples where whole 'gardens' seem to have been preserved, with the arms frozen, as it were, in motion, and even the roots at the base of the stems in place.

The crinoids have had an eventful geological history. True crinoids are doubtfully known before the Ordovician, but once established they diversified rapidly in the manner we have seen repeatedly with other groups. They soon spread to a variety of habitats, but in the Palaeozoic they were conspicuously abundant in relatively shallow environments. The most mouth-watering specimens tend to be found in limestones – famous sites are in the Silurian rocks of England and Mississippian rocks of the USA. The group as a whole had a major crisis in the Permian, during which most of the Palaeozoic forms died out. A very few of these survived into the Triassic. In the Mesozoic there was another great radiation of the crinoids, the typical forms having flexible arms, and it is crinoids of this kind that survive today. Some of the most famous crinoidal deposits are of Jurassic age, and marvellous specimens have been collected from the Lias rocks of Britain and Germany. A new and successful innovation was the evolution of stemless crinoids – some of these acquired pelagic habits, and during the Cretaceous the genera *Marsupites* and *Uintacrinus* were widespread enough to be useful marker fossils. Some of these stemless forms developed grappling hooks at the base of the calyx   with these they can attach in a favourable site, and move when conditions become adverse. Such feather stars are very numerous in some reef habitats today. The long and varied history of the crinoids demonstrates how well the echinoderms have attacked the problems of filter-feeding.

ABOVE: Blastoid, *Pentremites robustus*, Carboniferous (Mississipian), Illinois, USA; 4 cm (1½ in) long. The specimen, viewed from the side, is preserved in its original calcite with only the slightest compression. In some localities blastoids like this one break easily out of the enclosing matrix, or are weathered out in considerable numbers.

LEFT: The Silurian cystoid, *Pseudocrinites magnificus*, Wenlock limestone, England; 5 cm (2 in) long. This specimen is preserved in a grey limestone-shale matrix. The fossil is composed of individual calcite plates, and has a stem made up of a number of rings. This rare and peculiar fossil looks rather like a sea-lily without arms.

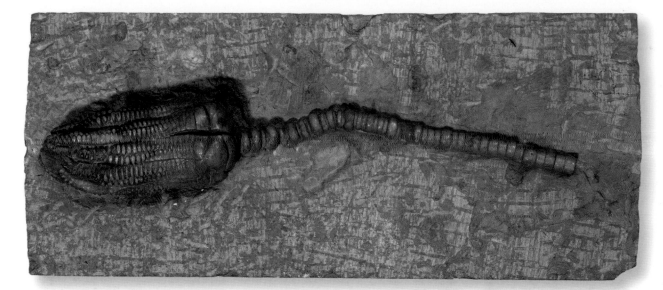

ABOVE: Triassic crinoid, *Encrinus liliiformis*, Muschelkalk, Germany; 18 cm (7 in) long. The calcite is preserved with a gloss on the surface in a matrix of limestone. *Encrinus* is typical of marine limestones of the Triassic, Europe.

BELOW: Carboniferous blastoid, *Pentremites spicatus*, Grayson Co., Kentucky, USA; 3 cm (1 in) across. Similar species occur in North and South America.

In the Palaeozoic rocks there were other stalked filter-feeding echinoderms that failed to survive the crisis at the end of the Permian, which was weathered by the crinoids. These extinct groups can exceed crinoids in number and variety at certain horizons, and they were evidently competing on equal terms. The blastoids (Class Blastoidea) had compact cups up to a few centimetres long, with five broad food grooves running down the sides (lacking crinoid arms). In life the food grooves were flanked by lines of delicate armlets (brachioles) that served to gather the food. Blastoids are sometimes abundant fossils in rocks (usually limestone) of Silurian to Early Permian age, and their perfect, compact pentameral symmetry makes their calices along the most attractive of fossils. Like crinoids they evidently grew in gardens, for their remains (especially in Carboniferous rocks) tend to occur packed together in thousands. Cystoids (Class Cystoidea) are even odder animals: often rather irregular bags of calcite plates, or if composed of a few plates these may carry powerful ribbing. The fivefold symmetry is often hard to detect in the calyx as a whole, although five food grooves are usually developed. They are a primitive, but very interesting, group with Cambrian origins, and reached their greatest variety in the Ordovician. Some of them evidently lacked a stem, and most have lain loosely on the ocean bottom. There are several more groups of odd echinoderms in the Lower Palaeozoic rocks – some of them only relatively recently discovered, like the bizarre helicoplacoids, which look like nothing so much as spinning tops. All are of the greatest interest, and any plated animal from the Cambrian is likely to be an important specimen. Sometimes these early forms are found as moulds – the original calcite having dissolved away. The external moulds then present a characteristic appearance, like the impression of a mosaic pressed into the rock. It is well worth keeping a special watch for fossils of this kind when hunting in Cambrian localities – it is still perfectly possible to discover a completely new kind of echinoderm.

Sea urchins – Class Echinoidea

Anyone who has waded into the clear blue waters of the Mediterranean without proper footwear may have encountered the protective covering of the sea urchins. Many of the living echinoids are protected by spines, some sharp and breaking off easily into an unwary foot, others stout and club-like; a few groups have taken to burrowing into sediment, and the spines have become small and felt-like. Under the spines there is a test composed of hundreds of calcite plates. Of course there is no stem – the sea urchins are self-propelled. Looking more closely at the arrangement of the calcite plates the fivefold symmetry is still in evidence – in this group there are five areas of finer plates (ambulacra) radiating from the centre. Under a lens these small plates show perforations, and during life the tiny tube feet passed through these to assist the animal in locomotion. The perfect polygonal joinery of all the plates is a striking feature of the echinoid test. Some sea urchins are almost spherical, with large strong plates looking rather like shields with a boss in the centre. These kinds of sea urchins carry the stoutest of spines attached to the bosses. Many living sea urchins have undergone a modification of the pentameral symmetry typical of the rest of the echinoderms – they have become bilateral again. Perhaps this is not surprising in animals that are accustomed to moving in a particular direction – they need a front and a back. In burrowing forms the mouth moves to one (forward) side and the anus to the back, which is obviously a sensible arrangement. Some species have become greatly flattened like the living sand dollar, which has a covering of very fine spines and can bury itself in sand with remarkable speed.

The sea urchins are among the most efficient scavengers in the sea – hence their somewhat unsavoury association with human effluvia. They also include species that eat

BELOW: Cidaroid sea urchin. *Plegiocidaris coronata*, Jurassic, Ulm, Germany; each 5 cm (2 in) across. Two beautiful examples of this sea urchin are illustrated, preserved in a fine-grained limestone in full relief. Note, however, that the stout spines are not preserved on these specimens. *Plegiocidaris* is found in Triassic and Jurassic rocks of Europe.

RIGHT: Jurassic sea urchin, *Hemicidaris intermedia*, Wiltshire, England. This specimen retained its spines, which are characteristic of this species, where they fell around the animal as it died. The matrix is a yellow limestone. The longest spine on this specimen has a length of 7 cm (2½ in).

RIGHT: Sea urchin, *Tylocidaris clavigera*, Cretaceous, Gravesend, England; 8 cm (3 in) across. Magnificently preserved in its original calcite in a matrix of chalk.

their way through enormous quantities of sediment, extracting edible particles from it, and in the process reprocessing all the sediment; some of the fossil heart urchins probably had this habit. These vacuum cleaners of the sea are vital in preventing fouling of the marine environment. Species with club-shaped spines include active hunters (although they tend to hunt sluggish animals like other sea urchins) and they have powerful jaws on the lower side of the animal that can munch their way through a sand dollar as if it were a biscuit. Some urchins use their jaws to make burrows in solid rock.

As with other major echinoderm groups, the geological record of the sea urchins goes back to the Ordovician. In my experience they are generally rare fossils in the Palaeozoic, but as so often with echinoderms there are localities where large numbers of specimens can be recovered from a bedding plane or two. They must have been gregarious from the start. The early sea urchins tend to have rather a large number of plates in a much less regular mosaic than their later relatives. By the end of the Palaeozoic, echinoids with club-shaped spines and beautifully regular tests had become well established, and their distant descendants survive today. In the Mesozoic a considerable proliferation of urchins occurred, and they acquired the importance in the marine economy that they retain. They are abundant in Jurassic and Cretaceous rocks, represented both by the spherical spiny forms and some of those with bilateral symmetry. Isolated spines are common fossils, and useful ones too, because their patterns vary from species to species. Some of the burrowing forms have left their burrows behind as well. In England sea urchins are especially easy to collect from the Cretaceous chalk, where they have been used as one of the fossils for dating the rocks.

ABOVE: Sea urchin, *Schizaster canatifera*, Pliocene, Perpignan, France; 8 cm (3 in) long. This specimen is perfectly preserved, but the fine fuzz of tiny spines that covered it in life are not present. This species looks like the heart urchin but the petal-like areas are more deeply sunken, the forward pair very short.

LEFT: Cassiduloid sea urchin, *Pygurus costatus*, Jurassic, England; 8 cm (3 in) across. The species is preserved without flattening in a hard limestone that fills the interior of the specimen. The five ambulacral areas – like the petals of a flower – are clearly visible. They are not sunken into hollows as on many echinoids .

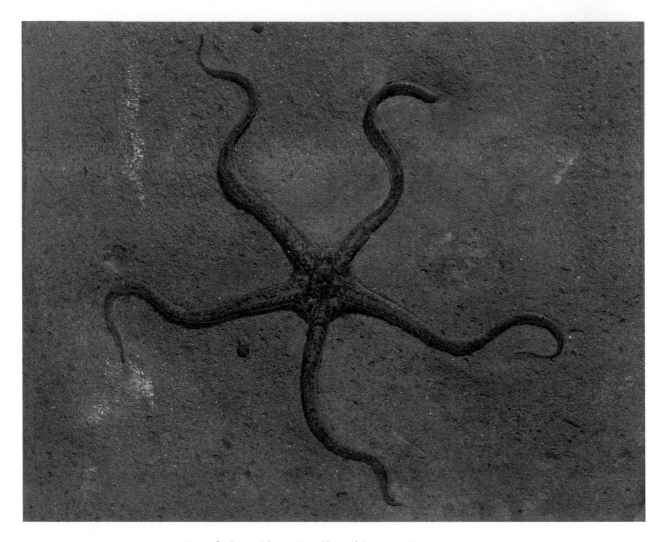

ABOVE: Fossil brittle (serpent) star, *Palaeocoma egertoni*, Jurassic, Lias, England; 8 cm (3 in) across. A beautiful species preserved in an impure limestone, deposited in very quiet water conditions, which accounts for the preservation of this delicate little fossil. The long slender arms are flexible. Brittle stars also occur in Miocene rocks in Maryland, USA.

Starfish – Class Stelleroidea

The starfish are among the most important of invertebrate predators living today. They make major depredations on oyster beds and the recent activities of the 'crown of thorns' starfish (*Acanthaster*) in chewing up great chunks of the Great Barrier Reef have become a modern ecological object lesson (and, incidentally, a source of funds for many marine biologists). There are two major kinds of starfish (which may not be closely related) – the familiar large, stiff-armed asteroids, and the delicate ophiuroids, brittle stars, with lithe, slender, snake-like arms radiating from a circular central disc. Ophiuroids are especially abundant in deep-sea environments, and it is unusual for a deep-sea dredge to miss one of these small animals. Asteroids often have five arms, but many species have more – up to about forty.

Although they are common today starfish are not generally common fossils, but (as so often with echinoderms) they seem to be local in occurrence. Their remains are commoner from the Mesozoic onwards, and at some levels, as in the Cretaceous chalk, they can be found in considerable numbers. Their history goes back at least to the Ordovician, although some of the Palaeozoic asteroids are distantly related to the living species. Ophiuroids have fossil representatives as old as Silurian.

LEFT: Fossil starfish, *Archastropecten cotteswoldiae*, Jurassic, Gloucestershire, England; 7 cm (2½ in) across. The specimen is in a fine-grained limestone, and the original calcite of the starfish is stained slightly yellow. This species has a small central disc, and long, slender arms. A related genus, *Astropecten*, occurs in Tertiary rocks, and survives today.

LEFT: Starfish, *Crateraster coombi*, Cretaceous, Kent, England; 10 cm (4 in) across. This starfish is preserved in soft white chalk. It has a broad central disc, and slender arms with stout plates around the perimeter. Similar forms occur in the chalk over Europe and in Cretaceous rocks in North America.

THE CARPOIDS

One of the oddest groups of animals covered with calcite plates are known as carpoids (occasionally referred to as calcichordates). Although they have been known for many years their evolutionary position has been hotly debated. They lack any pretence of fivefold symmetry: some are bilaterally symmetrical, but others have no obvious plane of symmetry at all, a most unusual thing for any animal. They appear to have a stem: could they be related to the crinoids? But the 'stem' tapers out to a point, and in some cases had a whip-like flexibility; the same structure has been interpreted as a 'tail'. Similar kinds of arguments can be applied to different parts of these extraordinary fossils. Internally they are really quite complicated, and some possess lobes and channels incised within the plates strikingly similar to the appearance of the brain and nerves in some chordates. Some authorities have suggested that the carpoids may include ancestors of particular chordates. They have a record going well back into the Cambrian, when it might be supposed that the chordates were undergoing a major diversification. On the other hand, recent discoveries in China suggest that the chordates had already appeared early in the Cambrian and it is unlikely that their ancestors should be known only from younger strata.

RIGHT: Carpoid, *Cothurnocystis evae*, Ordovician, Ayrshire, Scotland; 4 cm (1½ in) long. This curious species is preserved in a fine-grained sandstone as an external mould, i.e. all the original shell material has been dissolved away, but this still allows an accurate view of the original organism.

THE ARTHROPODS – PHYLUM ARTHROPODA

There are more living arthropod species than all other phyla combined; if diversity is a measure of success then the arthropods are the easy winners. Arthropods are those animals with an external, usually 'chitinous' skeleton (exoskeleton) and have characteristic legs, feelers, etc., with joints to give them flexibility (arthropod is derived from the Greek for 'jointed leg'). They have bodies divided into segments, and their legs and jaws are arranged in pairs. They are mostly small and there is a good physiological reason for this: gigantism is mechanically difficult for animals with external skeletons, and above a certain size it is difficult to get enough oxygen into the organism to sustain activity. Despite this one limitation the arthropods have the most varied habits and habitats of any phylum:

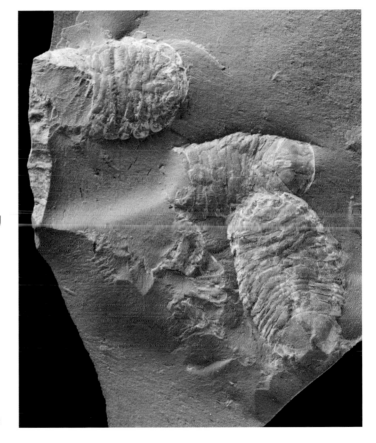

BELOW: Three individuals of the Cretaceous isopod *Archaeoniscus* from Wiltshire, England. Isopods are a large group of living arthropods, including woodlice and many subaqueous species, but they are uncommmon fossils.

they can be found anywhere on land, sea or in the air. They inhabit the driest parts of the desert and the deepest parts of the ocean with equal facility. This is partly due to the flexibility of the external skeleton – it can be modified into wings, fortified into claws, sealed against searing heat, or turned into eye lenses. Segmentation, too, has proved versatile; individual pairs of legs, or appendages, can be specialised for particular functions without affecting the routine business of locomotion taken by the others; appendages are modified for feeding in a host of ways, for grasping, swimming, spinning, copulating, cleaning or camouflaging. And their small size enables tiny arthropods, some almost too small to see with the naked eye, to live in crannies, within the soil, between sand grains, and the like. They are truly ubiquitous.

They are also an ancient group, since obvious arthropods, of primitive kinds, can be found among the earliest of Cambrian rocks. At this time, of course, they were wholly marine. With certain exceptions, the fossil record of the arthropods is not commensurate with their diversity today, and their probable diversity in the past. Some forms have delicate exterior skeletons that are only preserved under exceptional circumstances, as in amber. Many terrestrial forms simply did not live in environments in which preservation was possible. So the record of the group as a whole is a bit patchy, with 'jumps' in time between one fossil find and the next. Some groups, especially those with good, solid (and especially calcified) hard parts, are frequent and important fossils, with an unparalleled record. A brief summary of the arthropods cannot do justice to all the fascinating specialised groups – sea spiders and mites for example – that have an incomplete fossil record. We shall concentrate on the arthropods with the best fossil record. Some of these are common enough to dominate fossil assemblages, and they are among the most attractive of the common invertebrate groups.

Trilobites – Class Trilobita

The North American Indians had a name for trilobites that meant "little water bug in the rocks", which shows extraordinary zoological acuity. The trilobites were one of the dominant forms of marine life in the Palaeozoic, known from thousands of different species and found in every continent. Not surprisingly they were among the first fossils to attract widespread attention. They have a high degree of organisation and a variety of form that shows how quickly the arthropods were able to exploit the advantages of their external skeleton. All trilobites are divided lengthways into three lobes – hence their name. They can also be divided crossways into three regions, a well-defined head (cephalon) at the front, usually equipped with a pair of eyes, a thorax with a variable number of articulated segments, and at the back end a tail (or pygidium) formed by the fusion of several segments. The parts of the animal we find fossilised are usually only the carapace, the hard parts that formed a protective covering on the back of the animal. If we could have turned over a living trilobite we would have seen an array of jointed legs on the underside, and flexible antennae. The traces of these appendages are preserved in exceptional circumstances, so we do know that trilobites had walking legs, gills, antennae, and the bases of the legs modified into primitive jaws in some species. The appendage pairs are generally similar along the length of the animal. Few examples of complete trilobites are known and for the most part we have to be content with the exterior skeleton.

BELOW: Spiny odontopleurid trilobite, *Acidaspis coronata*, Silurian, Worcestershire, England; 2-3 cm (1 in) long. This spiny little trilobite is preserved in a fine-grained hard limestone. Trilobites of *Acidaspis* type have thoracic segments extended into long spines, which also fringe the head and tail.

LEFT: Cambrian trilobite, *Elrathia kingii*; ½–3 cm (1/5–1 in) long. The original shell has been thickened by mineral growth to give the fossil a distinctive, medallion-like appearance. It has a large number of wide thoracic segments, and the tail is wide and well segmented. This species occurs in great numbers in the Wheeler Shale, Utah, USA.

LEFT: Cambrian trilobite, *Wanneria walcottana*, preserved on a fine sandstone. This is one of the most primitive trilobites known, with crescent-shaped eyes attached to the middle part of the head. This specimen is 10 cm (4 in) long.

The great variety of form of the hard parts surely shows that trilobites were adapted to a wide range of habitats and lifestyles. We know that they could live in extremely shallow to very deep-water environments, that most lived on or near the sea bottom, but a few were adapted to open-ocean swimming. They probably had a variety of different habits; some no doubt were scavengers, others sediment grazers, and it seems probable that some trilobites were also active hunters, pursuing worms and other soft-bodied prey, which they shredded using the spines on the base of their legs. The eyes of trilobites include some very sophisticated structures that show they were far from primitive visually. Many species have thousands of lenses – like a fly's eye – which were very sensitive to detecting any movements in their field of view. Others had fewer lenses of remarkably efficient construction; photographs have actually been taken (p. 145) using these ancient lenses in a camera – and they still focus precisely after 400 million years or so. The lenses, like the rest of the exoskeleton, are made of calcite.

Trilobites vary in length from a millimetre or two to 50 centimetres (20 inches) or more – the 'average' trilobite is 5 centimetres (2 inches) or so long. It is actually rather unusual to find a complete trilobite as a fossil – usually they are in pieces. Trilobites, like most arthropods, grew by moulting (casting off the old carapace and growing a new, larger one) and many of the bits we find are probably the 'cast offs' and not the dead animal. During moulting the head split into several pieces along special sutures, the thorax often parted into its individual segments and the tail was displaced. The early moults, 'baby trilobites', have been discovered. Some trilobites lost their eyes – probably those that burrowed in mud or lived in lightless parts of the ocean. Many trilobites could roll up when threatened, just like the living woodlouse – some of them even evolved locking devices to make their enrollment really secure. In some places you can find dozens of such trilobites together; these are the remains of the animals themselves, not the moults, which presumably perished together after a fruitless attempt to protect themselves from a miniature catastrophe such as a sudden influx of sediment.

Typical trilobites of several kinds are present in almost the earliest Cambrian rocks; they must have had Late Precambrian ancestors, but of these nothing is known. They swarmed in the Cambrian seas, outnumbering all other invertebrates in most localities. The trilobites evolved rapidly, and can therefore be used extensively in the dating of rocks at this time. They continued to proliferate in the Ordovician, although many of the Cambrian kinds had died out; in terms of the variety of forms present this period was perhaps their heyday. Several kinds of trilobite failed to survive the end of the Ordovician, but other trilobites were still conspicuous during the Silurian and Devonian. Only a few major groups survived into the Carboniferous, but they can still be locally abundant. The last trilobites are found in Permian rocks, and

BELOW: Rolled-up Devonian trilobite, *Phacops rana*, Ontario, Canada; 3-8 cm (1-3 in) long. This species is beautifully preserved in limestone in full relief. A convex trilobite with eleven thoracic segments. The mid-part of the head is covered with coarse tubercles and expands forwards; the eyes include a few, very large lenses.

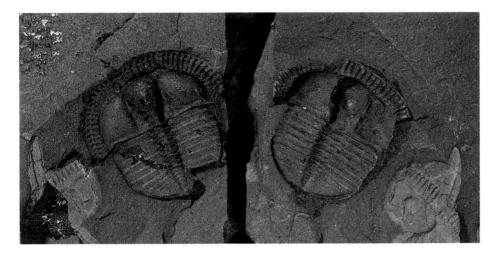

LEFT: Trinucleid trilobite, *Trinucleus fimbriatus*, Ordovician, Wales; 2½ cm (1 in) long. This specimen is preserved in shale with some of its exoskeleton coloured by iron compounds. Trinucleid trilobites are blind, and have a greatly inflated mid-part of the head region, with a pitted fringe around the perimeter of the head unique to this kind of arthropod. Part and counterpart are shown.

before the Triassic they had disappeared forever. We might perhaps hope that one species lingered on, like the brachiopod *Lingula*, in some inaccessible part of the sea, but as even the deep-sea faunas become well known this possibility is fading.

Since many of the extremely ancient rocks in which trilobites can be found are folded or otherwise distorted their remains can often be found twisted or fractured. But in other sites, particularly in limestones, they can be found with every fine skeletal detail preserved. The many different kinds are distinguished by: the nature of the segmentation, particularly in the head region; the structure of the thorax and pygidium; the size, position and structure of the eyes; the development of the moulting sutures; and the surface patterns on the exterior of the exoskeleton. New trilobites are being discovered every year, and there is always the possibility of finding one with its legs preserved.

BELOW: Asaphid trilobite, *Isotelus gigas*, Ordovician; 9 cm (3½ in) long. A 'smoothed out' trilobite, with the head a similar size and shape to the tail, only eight thoracic segments, and small eyes preserved in limestone. *Isotelus* is typical of the North American continent, where it occurs in many localities.

Crabs, lobsters, shrimps and their allies – Class Crustacea

The seas today swarm with crustacean arthropods ranging in size from the tiniest krill in the surface waters to giant bottom-living crabs with a reach like that of a human. The majority of Crustacea have appendages that are specialised to do different jobs along the length of the animal – some for grasping, some for swimming and so on. Most smaller Crustacea have exoskeletons that are rather delicate, and for this reason the fossil record of the shrimps, for example, is somewhat incomplete. Undoubted crustaceans are found in rocks as old as Cambrian, at which time free-swimming species were living happily alongside the trilobites. It is not until the Carboniferous that we start to find evidence of a variety of forms including the putative ancestors of some of the living dominant groups. Crabs and lobsters have the thickest exoskeleton of all the group – one like that of the trilobites reinforced with calcium carbonate – and the fossil record of this important group (only a small part of the whole class however) is much better. Lobster-like animals are found as far back as the Triassic, and both crabs and lobsters are frequent and appealing fossils in Cretaceous and Tertiary rocks. The group was hardly set back at all by the major extinction at the end of the Cretaceous, and has never been more varied than it is today. The most highly specialised crustaceans are probably the barnacles (Cirripedia) and individual barnacle plates are rather commonly encountered in rocks of Cretaceous age and younger. Finally the diminutive, bivalved ostracods are important and abundant fossils, which we will consider with other microfossils in Chapter 10.

ABOVE: Fossil crab, *Palaeocarpilius aquilinus*, Miocene, Libya; 6 cm (2½ in) across. This perfect and complete crab is exceptional, taking its colour from the white limestone matrix. It has a wide carapace, a spiny margin, and powerful pincers held close to the body, equipped with a good 'cutting' edge.

RIGHT: Fossil swimming crab. *Notopocorystes broderipi*, Cretaceous, Folkestone, England; 2½ cm (1 in) long. This crab is preserved in a soft clay making it rather fragile. It has a long body and the legs are flattened and held away from the carapace.

Spiders, scorpions, sea scorpions, horseshoe crabs – Class Chelicerata

This large and varied group of arthropods includes the giants of the phylum, as well as tiny spiders hardly visible to the unaided eye. They are classified together because of basic similarities in the appendages, even though they have diversified into very different-looking creatures living equally happily on land and in the sea. Predatory habits are typical, and many species have specialised clutching appendages, or stinging organs. None of the group is especially common as fossils, but they make up for their general rarity by their interest. Most spectacular are the sea scorpions (eurypterids), which include arthropods as large as any that have lived (2 metres, 6 ½ feet, or more in length). Appendages in the 'head' region often include powerful claws, and it does not take much imagination to conclude that these

ABOVE: Fossil lobster, *Thalassina anomala*, Pleistocene, eastern Australia; 12 cm (4½ in) long. The original shiny cuticle of the lobster is preserved in a yellow, limy matrix. All the legs are attached, some partly concealed. This species has long thorny pincers with little upturned 'fingers' at the end.

LEFT: Sea scorpion (eurypterid) *Eurypterus lacustris*, Devonian. The sea scorpion is preserved on a fine siltstone, and is exceptionally complete. Fragments are more usual. The rather fat body, with distinct segmentation, and a long tail-spine are characteristic; the head region carries appendages of grasping aspect. This species is from Canada, but eurypterids occur on most continents.

eurypterids were among the fiercest predators of their day. They were prominent in the Silurian and Devonian, where they can be found with the fish-like animals that abounded in the freshwater and brackish-water deposits of the time, although some of the Ordovician ones are in marine sediments. Some of the eurypterids may even have made tentative excursions on to the land. They are also probably the earliest animals in which two sexes can be certainly identified. The eurypterids are likely to be allied to the scorpions, which originated in the Silurian or even earlier, and successfully made the transition from the aqueous environment to land. They are abundant enough today in warmer climates to be a serious cause of death in some countries. Perhaps it was the transition to land that enabled the scorpions to survive beyond the Palaeozoic, which saw the end of the sea scorpions. Spiders are known as far back as the Carboniferous (spider relatives as early as the Devonian), but their remains are principally known fossilised from the Tertiary ambers, where perfectly preserved specimens retain even the hairs on the legs. The horseshoe crab (*Limulus* and its relatives) is one of the most celebrated of 'living fossils', and it is a genuine survivor having changed little in 150 million years. They are the last of the merostomes – the group of fossil

OPPOSITE: Horseshoe crab, *Mesolimulus ornatus*, Jurassic, Solnhofen, Germany; 17 cm (6½ in) long. This specimen is preserved on a flat-bedded limestone, and some of the original flexible material of the skeleton remains. These crabs have a nearly circular carapace, beneath which powerful legs helped the animal to swim and catch prey.

LEFT: Spider, *Abliguritor niger*, Oligocene; less than 1 cm (½ in) long. This spider is encased in amber, the colour of which is related in part to its age – darker amber is usually older. It is perfectly preserved except for its soft parts. This fossil is known only from Baltic amber.

horseshoe crabs that were varied and numerous in the coal swamps of the Carboniferous and have a history that extends back to the Cambrian. Why *Limulus* and its relatives should have survived is something of a mystery but they do have exceptional abilities to survive damage. It appears to have rather generalised habits, feeding on everything from worms to clams, which it can crush with its powerful appendages, using the bases of its legs like nutcrackers. It can swim, flipped over on to its boat-like carapace, or crawl. Presumably its ancestors were not so different when they shared the marine habitat with the trilobites.

Insects – Class Insecta

The insects – flies, butterflies, beetles, ants, fleas, bugs, dragonflies, grasshoppers, aphids, wasps, etc. – are the most varied and numerous of the arthropods both in numbers of species and individuals. They are also the most perfectly adapted of terrestrial organisms, inhibited neither by extremes of climate nor altitude. If this book were written in proportion to the number of species in the group the rest of it would have to be devoted to the insects. But the fossil record of the insects is far from perfect – very sporadic and selective. Where insects are preserved they seem to occur in large numbers and in variety, so that a few famous localities have yielded a disproportionate number of the fossil species. Insects are fragile and easily damaged, and it is not surprising that they are sometimes destroyed completely. Many species live, and presumably lived in the past, in inland or upland sites where little sediment accumulates. Naturally the commonest insect fossils are the most durable parts, like the wing cases (elytra) of the beetles. But their small size also means that insect fossils are easily overlooked and they may be more common than is supposed. The best sites are in rocks originating from sediments that accumulated in quiet, freshwater sites – lake bottoms, peat bogs and the like – and they are generally rarer in marine facies.

Most insects have six legs in three pairs, and are equipped with wings (a number of primitive forms are wingless). The early evolutionary stages of development of wings are

BELOW: Ant, *Iridomyrmex geinitzi*, Oligocene, Baltic amber; 4 mm (1/5 in) long. Ants frequently got caught in the pine resins that were destined to become amber, and numerous fossil species have been recognized.

Dragonfly, *Cymatophlebia longialata*, Jurassic, Solnhofen, Germany; wing is 7 cm (2 ½ in) long. The dragonfly is splendidly preserved on the flat bedding plane of a limestone. Much of the fine detail is preserved, especially on the wings. Dragonflies have long wings with finely netted veining, long slender bodies, and large eyes.

still in question – they may have developed from outgrowths of exoskeleton from the body that assisted gliding initially, and then acquired a propulsive function. In any case winged insects were well established by the Carboniferous, when some of the giant 'dragonflies' (not particularly a close relative of the living species) had a wingspan of tens of centimetres.

No doubt many of the crucial steps in insect evolution took place during the Devonian, but of this phase of insect history there is little record. In the Carboniferous and Permian the insects were highly diverse, and there is still debate over how closely these forms are related to our living fauna: some forms, like the cockroaches, undoubtedly have a long history. Flight gave the insects the chance to exploit habitats previously vacant, and unrivalled dispersal. The extraordinary flexibility of the group has meant singular problems in their classification, but certain characteristics, like the structure of the mouthparts or the wing vein patterns, serve to define major subdivisions. During the Mesozoic some of the most important types of insects still living – like the butterflies – originated. Butterflies have an intimate relationship with flowering plants, which they pollinate, and themselves derive nourishment from the nectar. It is difficult to imagine a butterfly without a plant to feed on, and we have to assume that both the plants and the appropriate pollinators diversified together, an example of co-evolution. Other kinds of insects besides butterflies have used the nectar of plants for a livelihood, including bees and some flies.

The fossil record of insects is at its best from Tertiary deposits, where amber, the fossil resin of trees in which insects became entrapped, yields great numbers of exquisitely preserved specimens. Even the most delicate are present. The oldest amber is Cretaceous. The fossil remains of beetles, including the hard elytra, have been extensively used in the interpretation of the climatic history of the Pleistocene ice advances and retreats, because the species of beetle present in peat deposits change in harmony with the climatic fluctuations.

PHYLUM CHORDATA

The chordates include the dominant living animals, if you measure dominance by size, or the capacity to alter the world. Man is a chordate. The phylum at the present is dominated by its more advanced members – the vertebrates. But in the Cambrian or earlier the chordates no doubt included animals little more highly organised than the contemporary arthropods or brachiopods. A few of these survive. The lancelet, *Branchiostoma*, one of the standard laboratory animals for dissection, is a primitive chordate (there is a fossil lancelet from the Cambrian). The acquisition of a skeleton of calcium phosphate marked the point at which the fossil record began to contribute to the history of the group which, from our selfish viewpoint, we are conditioned to think of leading to the pinnacle of the evolutionary tree. All chordates have a nerve chord running along the back, and all higher chordates were derived from forms with gill slits. We are mostly concerned with the history of the vertebrates. Once the vertebrates were established they quickly acquired hard parts displaying great complexity when compared with most invertebrate phyla (we should except the arthropods, perhaps). So it is possible to deduce more about the course of evolution from the study of vertebrate skeletal remains than is possible with most invertebrate groups.

As a generality, the fossil record of the vertebrates is not as complete as that of many of the invertebrate groups, and particularly those with a continuous fossil record like the ammonoids. Vertebrate remains are decidedly patchy, particularly the terrestrial species. This is not altogether surprising considering the factors militating against their preservation – break-up of skeletons after death, the necessity to have a skeleton incorporated into sediment, and the activities of predatory and scavenging animals to destroy remains. So land vertebrates in particular tend to come from relatively few sites, which become exhaustively collected compared with most invertebrate localities. Some of the sites are enormously rich, however, and the history of palaeontology is punctuated by quite unscientific feuds between experts trying to find and hold on to the best sites for the most spectacular vertebrates. The more impressive the animal the greater the feud; the fights between Professors Cope and Marsh in the 19th century over the dinosaur remains of North America include examples of double-dealing that would not shame an oil tycoon. Even fish remains are rather local, and do not adequately reflect the abundance of a group that has occupied the seas for more than 400 million years.

There is room here to give only the briefest account of the different kinds of vertebrates. The higher vertebrates, fossil mammals and dinosaurs in particular, have been described in so many books that they sometimes seem to have as much flesh and blood as any animal in the zoo. The dinosaurs alone have commanded as much popular attention as the rest of the fossil animal kingdom combined. This reflects their spectacular size and the drama of reconstructions of battles to the death between *Tyrannosaurus* and its armoured contemporaries. The sad truth is that the average collector is unlikely to recover more than the most fragmentary remains of dinosaurs. The commonest vertebrate remains are of teeth or other resistant pieces. Fortunately, the teeth of higher vertebrates can tell us a good deal.

Jawless fish

The oldest fossil fish, which are the earliest known vertebrates, lack jaws – the mouth is a simple opening – and are covered on the outside with bony skeleton. The first ones seem to have lived in the sea, but by the Silurian the jawless (agnathan) fish were a prominent

LEFT: A Devonian armoured fossil fish, *Botriolepis canadensis*, from Quebec, Canada. The trunk is 5 cm (2 in) long and covered in thick bony plates.

component of freshwater and brackish-water sites in many areas of the world, and continued to diversify into the Devonian period. Some of these freshwater fish are heavily armoured and make robust fossils – particularly the headshields. The insides of the shield have been known to preserve quite tiny details of the nervous system, for example. It is probable that the jawless fish lived by grubbing in the sediment, or perhaps by filter-feeding, possibly exploiting organic material derived from the plants that were taking to life on land at the same time. A few living jawless fish are the only remnant of this ancient group, and they are highly specialised forms. The hagfish feeds on dead or dying fish using its rasp-like tongue in conjunction with a solitary tooth on the roof of its mouth. The lampreys (*Petromyzon*) are rather nasty external parasites of other fish, to which they attach themselves with a sucker, and proceed to rasp away at the living flesh. The specialised habits of both these 'pseudo-fishes' may have contributed to their survival, when none of their bony-plated relatives survived the Palaeozoic. Modern specialists often divide the jawless fishes into two broad groups, one containing the hagfish and the other the lamprey, with the former generally considered the most primitive.

Jawless fish are particularly common in red rocks of the Silurian and Devonian periods, which was a time when non-marine rocks were widespread, and have survived subsequent erosion. They accompany the eurypterid arthropods (which may have preyed upon them), and we can visualise the rivers and lakes of the time thronging with the first invaders from the sea. The subsequent development of the jaw enabled the vertebrates to exploit a wider range of lifestyles than were available to the jawless fish, and it would not be too gross a generalisation to say that much of the evolution of the vertebrates was intimately related to things that happened to jaws.

Sharks and allies

The sharks and their relatives are the most successful of long survivors among the vertebrates. Their rapacious habits have been the subject of bloody hyperbole in films and bestsellers, so everybody knows that the shark has ranks of fearsome, pointed teeth, an elegant, incessant swishing swimming motion, a tail with a long 'point' uppermost, and a big appetite. Their skeleton is made of cartilage — bone has not 'grown into' the cartilage, as in the higher vertebrates. Their true jaws are of obvious advantage for grasping prey. Cartilage does not preserve as fossil as a rule, so most of the evidence of the shark-like fish rests upon teeth. Sharks of essentially modern type go back to the Jurassic. In the later Palaeozoic a number of shark-like fish include some related to living forms, and others whose zoological position is still not settled. The Devonian placoderms are an especially exciting group of fish with powerful, armour-plated heads and frequently fang-like teeth, the function of which is obviously predatory. The lower jaw of these fish is loosely slung to the brain case so that they had a very wide 'gape'. The trunk was also protected with a shield. This group may include the distant ancestors of the sharks, but they must have been somewhat ponderous animals compared with the streamlined hunters of modern seas. Most of them were dogfish-sized, but *Dunkleosteus* was twice the length of a human, and was surely the most formidable predator of its day.

Sharks of modern type replace their worn teeth with new ones by a sort of conveyor belt system. Isolated, bladed shark's teeth are really rather frequent finds in Mesozoic and Tertiary rocks, often in the absence of any other fish remains. They have a handsome shiny lustre, which makes them conspicuous from a distance. The rays are related to sharks, and are in essence sharks that have been 'squashed' flat in response to life as rather sluggish bottom dwellers. The mollusc-crushing teeth of rays, with the same lustre as shark's teeth, but formed as low, ridged cushions, can also be found in Mesozoic and younger rocks.

RIGHT: Fossil shark's teeth,
Hypotodus robusta, from the
Early Eocene, Kent, England. Both
the anterior teeth, which are tall
and slender and the lateral teeth,
which are triangular, are shown.

Bony fish – Class Osteichthyes

Every experience with an unfilleted kipper demonstrates the aptness of the general name of this group. The vast majority of living fish belong here (18,000 species or so), and they rival the insects in their multitude of adaptations and variety of external form. They can live in hot springs, or can be dredged from the deepest abyssal depths of the ocean; the latter are grotesque gargoyles that seem to belong in the paintings of Hieronymus Bosch. Some are rapid swimmers gathered together in silvery shoals, others sluggish bottom dwellers spending most of their time buried in sediment. Most have a characteristic arrangement of paired fins, but these can be lost, or fantastically modified to form fans or poisonous barbs. Most species are a few centimetres to a few tens of centimetres long. Their rapidity of evolution is well known: dozens of species have evolved in African freshwater lakes within the last million years. They have as wide a range of life habits as any group, from scavengers, grazers, filter-feeders, vegetarians and omnivores, to fierce predators with reputations to rival that of the sharks. A coating of scales is characteristic, and the gills are covered by a movable flap (the operculum).

It would be difficult to overestimate the importance of the bony fish in the economy of the sea. Fish of various sizes feed on the plankton (and on each other), and themselves are food for larger predators, such as seals and many whales, and humans. Many of them have bizarre adaptations, as parasites, or as permanent 'guests' within particular species of coral or sponge, which it would be impossible to infer from fossil remains. Many have good sight,

BELOW: This Jurassic *Dapedium* fish has thick rhomboid scales that closely interlock with one another and formed a tough protective cover to the body. The head bones are also very thick and provided further protection against predators.

but there are a few forms which have lost their eyes, some of them living in the lightless world of caves. Considering their abundance in the oceans and rivers today, bony fish fossils are rather rare, and poorly reflect their true importance. Where they occur, however, they are found in large numbers, and in variety. Presumably these sites were protected from the kinds of influence that normally destroy fish remains, such as the activity of scavenging animals that disarticulate the skeletons, or currents. The classic fish sites produce fossils of quite outstanding beauty (see p. 120), and considerable monetary value. The fish of the Eocene and later are clearly related to today's fauna, and some relatives of the living groups can be found in Mesozoic rocks. The classification of bony fish is extremely complicated, particularly fitting in the fossil forms, and there are many bones of contention (literally). The group as a whole goes back to the Devonian (?Silurian) period, and from rocks of this age predatory fish (*Palaeoniscus* and allies) have been recovered. Some living survivors of these early forms include the sturgeons and bichirs (the former have reverted to a cartilage skeleton). The surface of the body of many early fish was covered with an interlocking series of shiny scales ('ganoid'), which are still found in the bichir. The early fossil bony fish also have tails with the long blade uppermost, whereas the more advanced living bony fish have a tail that is symmetrical, and more effective in producing a horizontal thrust.

One important division of bony fishes includes the lungfish and the 'lobe-finned' fishes (including the coelacanth, the most famous of living fossils). This group of fish was also fully fledged in the Devonian, and this is of particular significance because the origin of all land vertebrates has been considered as lying within a species of this group. The freshwater lungfish retain their lungs, which help them to survive periods of drying in the rivers in which they live. Many fish specialists believe that in the Devonian all bony

ABOVE: Lungfish, *Dipterus valenciennesi*, Devonian, Achanarras fish bed, Caithness, Scotland; 18 cm (7 in) long. The specimen is flattened, but otherwise exceptionally complete on a bed of siltstone. It has a long body with overlapping scales, with a concentration of fins at the hind end.

BELOW: The early jawed fish, *Cheiracanthus murchisoni*, Devonian, Banffshire, Scotland; 15 cm (6 in) long. This fish is covered with minute rhomboidal scales, and prominent spines supported paired fins. In Devonian rocks, fishes of this kind are known from mostly freshwater deposits.

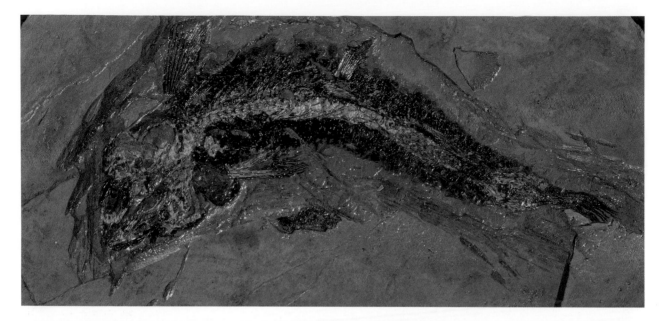

fishes had lungs (and that the swim bladder of the recent forms was a modification of the early breathing apparatus). It is not easy to identify the actual ancestor of the land-going organisms, but a recent discovery in Arctic Canada of a fish called *Tiktaalik* has revealed a very good candidate for an intermediate form (see p. 220). For many years there were proponents of a theory that one of the 'lobe-finned' fishes included the ancestor of the tetrapods; their stumpy fins, with a fleshy core, supposedly made the uneasy transition into a walking leg. The new discoveries favour *Tiktaalik* and its relatives as nearer the origin of amphibians and, ultimately, ourselves. Whatever momentous steps they took in the Devonian, the lungfish and lobe fins have been very conservative since, for the living lungfish are clearly similar to their Palaeozoic relatives and coelacanth 'lobe fins' like the living genus *Latimeria* are known from Cretaceous rocks.

Amphibians – Class Amphibia

Frogs, newts and salamanders are living amphibians. The frogs in particular are one of the most diverse of the higher vertebrate groups and peculiarly endearing animals with their lugubrious expressions and vocabulary of squeaks and belches. Yet the amphibians as a whole represent an early level of terrestrial organisation; they are tied to water for reproduction, because their larvae have to pass through a wholly subaqueous phase, and desiccation of the adult is a constant danger. The amphibians had their origins in one of the freshwater lung-carrying fishes in the Devonian. We have to visualise the process of change as a gradual one, with progressive forays from the water, until the first animal with true limbs developed (the bony elements of the limb are present in the fish fin). From the Devonian of Greenland and other localities comes a true four-legged animal (tetrapod) that is not far from the 'fish with legs' we might have visualised (which is commemorated in its name – *Ichthyostega*). But as always it is not quite that simple. The transition from water to land required not only the perfection of a walking limb, but modifications to the hearing system, palate, eyes and so on, and *Ichthyostega* had already established most of these terrestrial modifications, so it is in no sense an 'ancestor', although it has many ancestral features. Bony scales were inherited from its ancestors and no doubt this helped the animal against desiccation. The subsequent history of the amphibians includes a further development of terrestrial forms, but others remained wholly or partly aqueous. These flourished in the waterlogged swamps of the Carboniferous period. The earliest amphibians were certainly predators, and there was suitable prey both on land and in the water.

The Carboniferous is popularly known as the 'Age of Amphibians', and it is true that the amphibians reached a dominance in the vertebrate world at this time which they never again equalled. All of the walking species had the widely splayed legs that gave them a slow and lumbering gait, but, in the absence of more streamlined animals, they prospered. The group informally known as labyrinthodonts (from their characteristic labyrinth-ridged teeth) included some impressive animals almost as long as a human: perhaps the alligators of the time. Most of the fossils recovered are of wholly or partly aquatic amphibians, the sort that were easily preserved in the coal swamps that were characteristic of the Carboniferous non-marine environment. Other environments, particularly later in the Permian period, which was one of general 'drying out', probably included species that were more truly terrestrial in their habits. But even during the Carboniferous the first true reptiles had evolved, and this was the group that was to make itself wholly independent of the water. The living amphibians have turned the limitations of the aqueous connection to advantage. Frogs have their origins in the Triassic (they are a relatively recent amphibian innovation, and highly specialised). Their tadpoles can exploit bodies of water not excessively populated with competitors, and some are adapted to very restricted niches. The adult frogs have a different diet (worms, flies, etc.), and the species therefore get "the best of both worlds". Frogs throng in almost any site with high humidity and standing pools of water. The large amphibians of the Palaeozoic did not survive beyond the Permian, and so our inferences about their modes of life have to be made entirely from the bony fossils that survive. Palaeontologists recognise fossil Palaeozoic amphibians from their teeth, their flattened skulls and the arrangement of bones on the skulls. They are not generally common fossils, but there is always a chance of turning up a skeleton from the coal deposits, or finding a fossil frog (see p. 124) in the Mesozoic.

OPPOSITE ABOVE: Bony fish, *Lycoptera middendorfi*, Cretaceous, Turga, Nertchinsk, Siberia; 6 cm (2½ in) long. This little fish is preserved in soft, pale grey shale. It has a rather expanded, frog-like head behind which are prominent fins, and the slender flexible body is supported by a host of tiny ribs.

OPPOSITE BELOW: The coelacanth was widely accepted to have died out with the dinosaurs, and was only known from fossils, its odd-shaped tail, thick scales and bony plates covering its head, all signs of a very ancient creature. But in 1938 one was found off the coast of South Africa. Since then, a living colony of more than 300 has been found near the Comoros Islands, northeast Madagascar, and two individuals from another species in Indonesia.

Rana Pueyoi Nav
Mioceno
Libros (Teruel)

Reptiles – Class Reptilia

The living reptiles are almost all predators, and the vast majority are also small, the only giants remaining being the crocodiles, alligators and caymans (and one or two big lizards). (I should point out that some scientists do not regard reptiles as a wholly natural group of animals, especially since the group does not formally include the birds – which were of reptile descent.) We tend to think of the reptiles as somehow past their 'prime', but it would be more accurate to say that they had been displaced from the top jobs in nature, while more than holding their own in the shop floor. The warmer the climate the more they are in evidence, because the living reptiles are 'cold-blooded' (without their own internal heat regulator) and they cannot successfully live in very cold climates. Living snakes and lizards are highly varied, and manage to live in some environments (e.g. dry deserts) where the mammals are pushed to survive. Most living reptiles and all of the primitive ones lay eggs, and this amniote egg, from which perfectly formed baby reptiles hatch, marks the complete emancipation from the necessity of returning to water for reproduction. And a scaly skin solves the problem of drying out even in intense heat. Add to this the change in orientation of the legs in many reptiles, which can be used in an efficient running action, unlike the ungainly waddling of the amphibians, and it will be apparent why the reptiles were better adapted to terrestrial life

OPPOSITE: A fossil frog, *Rana* sp. This specimen originates from shale deposits in Spain. The skeletal anatomy is preserved in its entirety with an outline of the soft parts clearly visible. This specimen measures 12 cm (4½ in) long.

BELOW: A fossil skull that belonged to the parrot-beaked dinosaur, *Psittacosaurus*. It lived during the Cretaceous period around 120–100 million years ago.

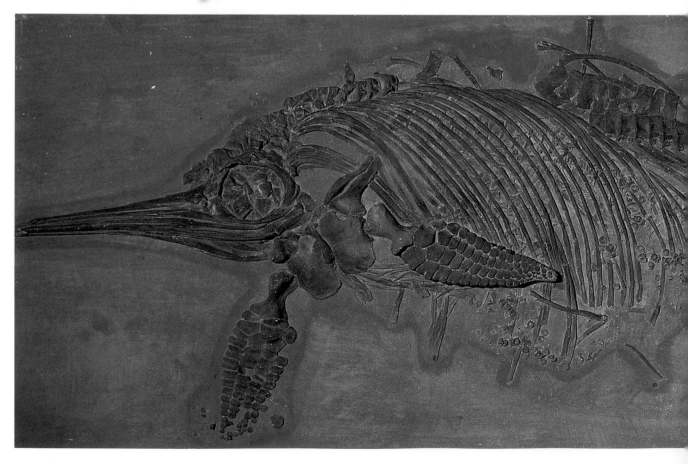

RIGHT: *Euoplocephalus* was a heavily armoured ankylosaur dinosaur with a huge tail club formed by two bony knobs fused together. It lived around 71 million years ago.

than the amphibians, and why they largely displaced them. This was far from instantaneous, and the Permian period saw both advanced amphibians and early reptiles living side by side.

The reptiles were probably derived from an amphibious ancestor early in the Carboniferous. Once established they underwent a number of evolutionary 'bursts' in which diverse kinds of reptiles occupied a variety of habitats, the most spectacular of which was the dinosaur radiation in the Mesozoic. By no means all of the large reptiles that are found in the Mesozoic rocks are dinosaurs — the reptile groups that took up life in the sea or the air were only distantly related. One of the most important steps for the complete colonisation of the land was the evolution of herbivorous (plant-eating) reptiles, tapping a prolific and rather nourishing source of food that opened up a new dimension to the 'food chain'. Larger and better herbivores meant larger and more ferocious carnivores to prey upon them, scavengers to clean up the mess, and a whole host of subordinate trades for other animals. The reptiles underwent tremendous changes in their skeletal structure, their jaws and their teeth during their history of more than 200 million years, some of the predators developing razor sharp rows of fangs, other herbivores losing their teeth and acquiring beak-like jaws. At some time in the Permo-Triassic one group of reptiles ('mammal-like reptiles') actually gave rise to the warm-blooded mammals, which were to lead a rather subordinate existence during the heyday of the dinosaurs. And the birds probably had their origin in a particular group of dinosaurs. So in a sense the reptile dominance is still with us, transmuted by time and evolution.

The scenes of Jurassic or Cretaceous landscapes in picture books show dinosaurs swarming over the landscape, giving the impression that we know the whole story. The truth is that there are a few sites (especially in North America) where abundant remains of complete animals have been recovered, and from which a passable idea of the fauna of the Mesozoic can be obtained, but there are still great gaps in our knowledge, and many of the dinosaurs are known from a handful of individuals, or even a single specimen. There are doubtless still new kinds to discover.

The dinosaurs have dominated the popular conception of the fossil reptiles, sometimes to the extent of seeming almost synonymous with the word 'fossil'! The group includes by far the largest of terrestrial animals ever to have lived (such as *Brachiosaurus*), creatures to dwarf the largest bull elephants. The predators that preyed on the giants were even more spectacular, and by now the name of *Tyrannosaurus* is so well known that it seems to be one of the first tongue-twisters mastered by small children. Of course there is little chance of an amateur collector acquiring an appreciable part of one of these great animals, but in the right site fragments of dinosaur skeletons may be picked up. The term 'dinosaur' includes a number of really rather separate groups of animals, with more or less independent

BELOW: The ichthyosaur, *Stenopterygius quadriscissus*. This ichthyosaur has been preserved with the broken-up skeletons of its unborn young inside. A fourth may have just been born; its skeleton can be seen below her tail. This specimen lived between 187 and 178 million years ago.

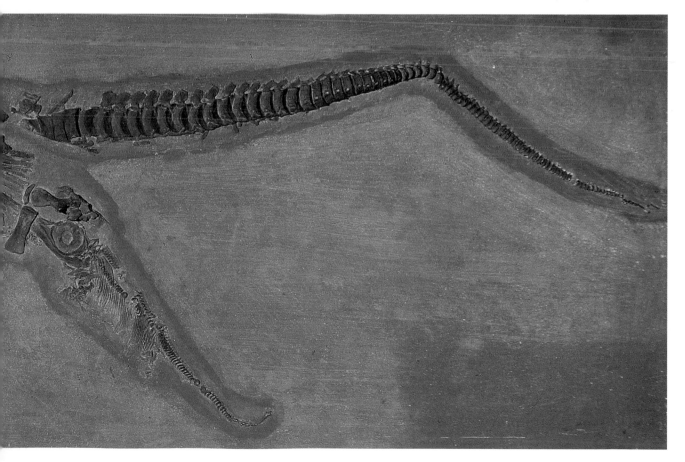

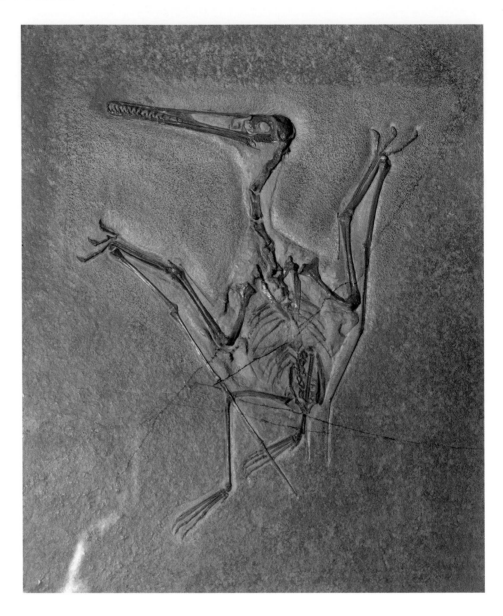

RIGHT: A fossil skeleton of *Pterodactyl kochi* discovered in the Solnhofen limestone of Bavaria, Germany. Its wingspan was about 46 cm (1 ft 6 in).

evolutionary history during the Jurassic and Cretaceous. The real giants were the lumbering sauropods (*Diplodocus*, *Brachiosaurus*, etc.), four-legged dinosaurs, with enormously long necks only matched by equally long tails. The giant predators that walked on their hind legs (*Tyrannosaurus*, *Allosaurus* and other theropods) share structures of the hip bones with the sauropods that show they are more closely related to them than to the rest of the dinosaurs. These latter (ornithischians) include some animals that walked on their hind legs like *Tyrannosaurus*, but with vegetarian habits (*Iguanodon*, the first dinosaur to be discovered); the group also includes various types of armoured and plated vegetarian dinosaurs, often shown in pitched battle with their carnivorous contemporaries. These well-protected dinosaurs each diversified into a number of genera. The three main groups are the stegosaurs of the Jurassic, with paired plates along the back and a nastily spiked tail (*Stegosaurus*); the ceratopsians, horned, rhinoceros-like dinosaurs of the Cretaceous, including the familiar

Triceratops; and the spiky armoured ankylosaurs, like tanks on stumpy legs. Specialised ornithischians in the Late Cretaceous included the remarkable duck-billed dinosaurs, animals that lost their front teeth and had arrays of tiny grinding teeth at the back of the jaw that were continually replaced, like those of the shark, and must have been able to cope with tough vegetation. Some of the duck-billed forms evolved crests and protuberances on top of their heads. All the dinosaurs, vegetarian and carnivore alike, became extinct at the end of the Cretaceous period (see Chapter 7).

A whole range of other reptiles were present in the Jurassic and Cretaceous; none of them are dinosaurs. They include the products of the spread of the reptiles back into the aqueous environment from which their distant ancestors emerged, and, for the first time in the history of the vertebrates, the conquest of the air. The marine plesiosaurs, with their long flexible necks and sharp teeth, were formidable hunters of Mesozoic fish. The limbs were modified into efficient paddles, perfectly adapted for sculling through the water. Even more streamlined for marine life were the ichthyosaurs ('fish lizards'), which, as their name implies, include species that look remarkably fish-like, although perhaps the better analogy would be with the porpoises, a group of mammals that 'returned to the sea', and may fill a similar role in modern seas to that of the ichthyosaurs in the Jurassic. These marine reptiles have a higher chance of preservation than many of the terrestrial species, and their carcasses often seem to have sunk to the sea floor, so that some localities have yielded numerous complete specimens. The Liassic (Lower Jurassic) rocks of Europe furnished many of the magnificent specimens that are now on display in museums. Some of the ichthyosaurs include baby individuals within their skeletons – which either implies cannibalistic habits, or, more probably, the ability to give birth to their young alive. Isolated vertebrae of marine reptiles are one of the reptilian fossils that are the easiest to collect.

The pterosaurs, flying reptiles, also had their origins in the Triassic and became widespread and varied during the Jurassic and Cretaceous. Like the bats, the pterosaurs had membraneous wings stretched between highly modified arms and 'hands' and the fore part of the legs, unlike the bats some of them grew to an enormous size and the front support for the wing was provided by only one finger of the hand, grotesquely extended. Most of them, and particularly the larger ones, combined slow, flapping flight with gliding, spending a large part of their lives effortlessly aloft. As in the case of the sauropod dinosaurs they seem to have got bigger and bigger during the Cretaceous – some of these later pterosaurs are supposed to have had wingspans exceeding 10 metres (33 feet), which would make them almost comparable to a man-made glider. Why this large size evolved is something of a mystery: it may have been that larger gliders cruise at slower speeds without 'stalling'. Quite how these amazing animals landed and took off (if they did) is still the subject of research. Like the plesiosaurs and ichthyosaurs, they were successful and varied for a long time, but all three groups failed to survive the Cretaceous.

Other groups of reptiles were not exterminated at this time, even though their fossils may be found in rocks as old as those that yield dinosaurs and the other spectacular, extinct groups. Crocodiles and lizards have a fossil record extending back to the Triassic, snakes to the Cretaceous. The turtles and tortoises have almost as long a history as the crocodiles. The fused, bony plates that protect their soft parts make them well-nigh invulnerable. All the long survivors show every sign of flourishing today: one might suppose that the tortoise will continue its lumbering progress no matter what catastrophes may come to pass in the future.

Mammals – Class Mammalia

Mammals – animals with mammary glands that suckle their young – were certainly present during the Triassic, and were a continuous if subordinate component of fossil faunas throughout the acme of the dinosaurs. It was only after the demise of the ruling reptiles that the class had the opportunity to capitalise on the evolutionary advantages that propelled them to the dominance they have enjoyed ever since. Many of the early mammals were probably insect eaters, as are many of the primitive representatives of the group today. Small mammals were present in the Triassic – shrew-like animals with long noses, and probably insatiable appetites for smaller items of the fauna. We can imagine animals like these darting through the undergrowth in search of food while the colossal reptiles lumbered obliviously around them. It would be wrong to suppose that there was no evolutionary activity among these early mammals: in fact there were considerable changes, which by the Late Cretaceous had established early members of several of the principal groups of mammals which dominate the terrestrial world today. The marsupial (pouched) and placental (womb-bearing) mammals had separated, and lived side by side.

Many of the early mammalian remains are fossil teeth. More than those of any other vertebrate group, the teeth of mammals have changed in distinctive ways, and the different types of mammals can be reliably identified from a study of their dentition alone. This is particularly fortunate, because teeth have a high fossilisation potential, exceeding that of other parts of the skeleton. The sediments filling caves or fissures in limestone are frequently repositories for fossil teeth. Sampling for mammal fossils often involves the patient sieving of great quantities of sediment to extract the fossil teeth. If you are very lucky you might find a whole jaw. Mammal jaws often have specialised teeth for particular jobs – nipping, chewing,

BELOW: Lower jaw of Pleistocene bear showing typical mammalian specialization of teeth: sharp canines with chewing teeth at the rear.

biting, gnawing teeth and so on – which function in perfect cooperation between the upper and lower jaws. Mammal teeth often have distinctive patterns of crests or bumps (cusps) that serve to identify large groups within the mammals. Teeth for grinding vegetation are often large and ridged, while the rodents have their fast-growing gnawing teeth at the front of the jaw. The most specialised dental structures of all are probably the baleen plates of the largest whales, modified for straining plankton from sea water.

The evolution of mammals after the Cretaceous was both extremely rapid and very complex, and we can do no more than give the roughest sketch here. Compared with what was happening in contemporary marine invertebrates, the visible changes in mammalian skeletons during the early part of the Tertiary were fast and dramatic; it did not take long before the roles of larger herbivores and carnivores were filled by suckling animals. By the end of the Paleocene, only a few million years from the Cretaceous extinction of the giant reptiles, there were representatives of many of the living mammalian orders, including, for example, the primates (the order to which humans belong), the carnivores (cats, dogs, and most living predators), and the rodents (rats, mice). The history of the mammals in the Tertiary involved not only this one major diversification, but several, with major extinctions

BELOW: *Glyptodon clavipes* was the most armoured of all the ice age mammals. This giant relative of the modern armadillo grew to about 3 m (9 ft 10in) from head to tail. Its most distinctive feature was its shell, made up of about 2,000 small plates of bone, each about 3 cm (1¼ in) thick and fused together to form a spectacularly off-putting defence.

between, so that the largest mammals of the earlier parts of the Tertiary are not related to the large mammals of the Recent. The roots of our present European fauna extend back to the Miocene. The whole story is complicated by the fact that this was also the time of break-up of the supercontinent Pangaea. Mammalian faunas on isolated continental fragments could evolve, at least for a time, separately from faunas in other parts of the world, producing a whole series of peculiar animals that have no direct relationships to animals with a similar mode of life elsewhere. The obvious example is Australia, which was separated early, and where the marsupial mammals had the opportunity to adapt to a whole range of ecological niches, which they managed with remarkable success in spite of a low cranial capacity and a primitive mode of reproduction. Some of the largest marsupials (*Diprotodon*) seem to have survived to a time tantalisingly close to the present. Similarly, in South America a whole range of peculiar mammals evolved largely in isolation, including giant sloths, specialised grazing species, and the extraordinary glyptodonts. A lot of argument among mammalian palaeontologists seems to centre on how and when links between North America, Europe and Asia existed, resulting in interchange and invasions of animals between the two regions. It is clear that some groups of animals, like the proboscideans (elephants and allies), are much reduced now compared with the Late Tertiary, both in variety and geographic distribution.

A late stimulus for the evolution of distinctive mammals was the Pleistocene Ice Age, although a variety also became extinct at this time. A number of species, adapted to the widespread cold conditions, roamed widely around the northern hemisphere, including woolly mammoths and rhinoceros, and extinct species of bears. On the other hand, interglacial periods were very much warmer than conditions in temperate latitudes today, and these climatic fluctuations resulted in periods when hippopotamus and tropical rhinoceros thrived in the areas occupied by Europe's capital cities today. With the final retreat of the ice many of the cold climate specialists perished (possibly with the help of humans), although some of the species that accompany them, like lemmings or musk ox, survive today in the harsh conditions of the arctic tundra.

Birds – Class Aves

Fossil birds are usually distinctly rare, and to find a well-preserved and complete early example is a palaeontological event. The evolution of flight led to generally light and fragile skeletons, and it is perhaps not surprising that avian fossils are so uncommon and, when found, so often fragmentary. The derivation of the birds from a reptilian (dinosaur) ancestor is certain, and because the earliest fossil bird, *Archaeopteryx*, is Jurassic in age, it is usually assumed that the split from the reptiles occurred in the earlier part of the period. Unfortunately, there is much that is not understood about the history of the birds between *Archaeopteryx* and the good assemblages of fossil birds undoubtedly related to living species in the Eocene and Miocene. Cretaceous birds are generally scarce, and some of the 'toothed' birds that are known, like the flightless, aquatic *Hesperornis*, seem to be difficult to relate to any of the living species. It is probable that, like the mammals, the roots of the great variety of modern birds are to be found in the Cretaceous, but fossils, which could document this, are only now slowly coming to light. Some of the largest flightless birds went extinct in historical times. The story of the dodo on Mauritius is a familiar one, but it is likely that human beings were also responsible for the extinction of the giant moa in New Zealand, skeletons of which are still abundant in the remote islands.

FOSSIL PLANTS

The fossils of plants are some of the most attractive that the amateur collector is likely to find. Those plants that concern us here were terrestrial or lived in bogs, rivers or lakes. Like animals that dwelt on land, the terrestrial flora also arose from ancestors that lived in the seas, a change that is known to have happened before the end of the Silurian. However, marine algae were very important in the early history of the evolution of the Earth, and they are described in some detail later. They are not generally conspicuous fossils in the field. A few algae, however, produced hard calcareous skeletons, and were abundant enough to produce small gardens beneath the sea. Such calcareous algae have a long history, extending back into the Precambrian, and they are still numerous today. They were often components of reefs, and in some environments it is the algae that bear the full brunt of the attack by the breakers on the exposed, seaward side of the reef. They are also important rock builders; in northerly latitudes today limestones are being formed from the skeletal debris of such calcareous algae as *Lithothamnion*.

The first land plants must have been derived from a marine alga. We have to visualise this as a very gradual process, the early transitional plants perhaps creeping across mudflats,

RIGHT: Oligocene sumac tree, *Rhus stellariaefolia*, Florissant beds, Colorado, USA; 12 cm (4½ in) long. Long, graceful leaves are divided into about twelve spindle-shaped leaflets, and a terminal leaflet with a prominent vein in the middle. Living sumacs are familiar in temperate regions, their leaves turning a beautiful dark red late in the year.

or partially submerged. The change involved the production of a protective 'skin' to ensure that water loss to the atmosphere was not excessive, and the stems had to acquire sufficient rigidity to stand up without the support of surrounding water, but also the cellular structure had to allow the passage of nutrients to the growing shoots (ultimately, a vascular system). The evolution of such a plant could not have been achieved at a single stroke. Reproduction in these early plants was by means of minute spores which are recognised as early as Ordovician. Quite a variety of Silurian plants are now known, and by the end of the Devonian it is apparent that most of the problems of terrestrial living had been solved, to the extent that large tree 'ferns' of the time would have had dimensions comparable with forest trees today.

The fossil record of plants is not good enough simply to 'read out' the story of their evolution by collecting from successively younger, non-marine rock formations. Some of the earliest plants do not appear to be the most primitive, and vice versa. True, there is a general drift through time towards more complex plants, more perfectly adapted to life away from water, but primitive species have persisted alongside the innovations. The fossil record is full of curious plants that do not fit comfortably into classifications – this is one of the things that makes the study of fossil plants so fascinating. Consequently, there are arguments among botanists about the relationships of many plants, living and fossil, and some classifications have quite large categories containing only a few plants, which may be rare. Not all of the major groups will be described here, but only those that are likely to be found by the non-specialist collector. Some important groups, like the mosses and liverworts, have a sporadic fossil record extending back to the Carboniferous, but they are not likely to be encountered without a special search.

Plant fossils are often to be found in particular beds, reflecting conditions of deposition that were just right for their preservation. Of course, the rocks that contain abundant plant remains are mostly those that were deposited in freshwater lakes, or on deltas, or the fossilised remains of peat beds. There were probably plants living in hilly regions from quite early that have left little fossil record, and this may account for some of the gaps in the preserved fossils. Many fossil plants are now more or less black, carbonised ('coalified') compressions. In some cases, however, the fine structure, even down to individual cells, is splendidly preserved, and these are the most important specimens from a scientific point of view. The problems of associating the different organs of the same fossil plant species were mentioned earlier in this book (p. 25).

Early land plants – Division Psilophyta

The earliest land plants, to be found in Silurian and Devonian rocks, had simple shoots that arose from a creeping 'axis', which were little different in structure from the shoots themselves. Leaves, if they were present at all, were humble scales or spine-like projections. The upright twigs often branched in a simple way, forking into two, and then into two again, and sometimes terminating in little capsules that carried the spores. The plants are often preserved as dark markings on shales, and were less than half a metre in length. The best-preserved examples, from which their cellular construction has been described, are from the Devonian Rhynie Chert.

Ferns – Division Pterophyta

The feathery and delicate fronds of the ferns make them among the most beautiful of the foliage plants, and they attract a fanatical group of devotees dedicated to growing exotic species in inhospitable cities. They are also a very ancient group, with many Carboniferous representatives, and a number of 'pre-ferns' in the Devonian as well. Then as now they generally preferred humid environments in which to live. They reproduce by means of minute spores, which form in clusters on the undersides of the fronds, or sometimes on specially modified fronds. To be really sure that one is dealing with a fossil fern it should be possible to see the spore cases, because other kinds of plants

BELOW: Fern, *Onychiopsis mantelli*, Cretaceous, Sussex, England; 12 cm (4½ in) long. This delicate fern is preserved in very fine-grained sandstone. Slender and delicate branching patterns with many subdivisions are characteristic of this kind of fern. Related forms occur in North America.

Cycad-like plants

In the forests of the Mesozoic there were large numbers of trees and shrubby plants that looked at first glance like palm trees, with leafless stems crowned with bouquets of stiff, large leaves, deeply divided into long, unbranched leaflets. These were cycads and bennettites. A small number of cycad genera survive in the tropical and subtropical regions today, but they are much less conspicuous today than they were 120 Ma ago. The resemblance to palms is no indication of their true relationships, because the cycad-like plants were not true flowering plants like palms. They do, however, have a most extraordinary fructification, a large knob arising from the centre of the crown, looking something like a corncob, and bearing numerous, very large seeds. Nowadays, botanists divide the cycad-like plants into two major groups that may not be particularly closely related: the Cycadales, which include the living species, and the Bennettitales, an important Mesozoic group of plants that have smaller fructifications scattered among the bases of the leaves. Marie Stopes, who later became a famous pioneer of birth control, started life as a palaeobotanist interested in the cycads. Some specimens of the trunks of bennettites are among the most spectacular of fossils, and they even attracted the attention of Etruscans, who included them among their sacred relics more than four thousand years ago! The amateur collector is most likely to come across the cycad foliage, with its characteristic narrow central rib and flat leaflets. Their fossils are commonest in Triassic, Jurassic and Cretaceous rocks.

ABOVE: A trunk of the extinct Bennettitalean, *Cycadeoidea microphylla*, 22 cm (8½ in) wide, from the Jurassic of the Isle of Portland, Dorset, England. To 19th century miners of Portland Stone, the barrel-shaped trunk of these cycad-like plants resembled giant fossil birds' nests.

RIGHT: Frond of a cycad, *Nilssonia kendalli*, Cretaceous, Yorkshire, England; 15 cm (6 in) long. This unusually complete fossil is preserved in fine-grained siltstone. Cycad leaves have characteristic flattened leaflets arranged in rows on a very slender axis. Other well-known cycad localities are the Black Hills of Dakota and from Maryland, USA.

Conifers –Division Coniferophyta

The conifers are as varied and successful a group of plants today as they have ever been. But again they are an ancient group with ancestors back in the Carboniferous forest, a melting pot for plant evolution. Botanists are not fully in accord about how many separate groups of plants there are within the conifers, and they are all lumped together here. The fir cones of modern species will be familiar to the reader, often littering forests in their thousands, and containing seeds, which may be equipped with wings. The close, needle foliage of most modern conifers, which seems as much immune to desert heat as arctic cold, was probably derived from lusher leaves in the Upper Palaeozoic. In the Carboniferous swamps large trees of the genus *Cordaites* grew to 30 metres (100 feet) or more in height, with long slim trunks terminating in branches that bore strap-shaped leaves up to a metre in length. The cordaite 'cone' was a longer and more complex structure than that of any living conifer. The leaves of this tree are rather common fossils, striped with longitudinal veins that give it a superficial resemblance to the leaf of an iris. Most of the steps leading to the modern conifers seem to have taken place by the end of the Palaeozoic; no other group of organisms of such antiquity has retained such an unbroken hold over vast areas of the Earth as the conifers. Some primitive conifers, like the monkey-puzzle tree (*Araucaria*), have survived with little change since the Triassic.

The long-leafed plant *Glossopteris* is something of a botanical puzzle, but probably belongs within this broad group. It is common in later Palaeozoic rocks of South Africa, peninsular India, Australia and South America, and is thought to have lived almost exclusively in the cooler areas of the supercontinent Pangaea.

The ginkgos are often placed in a separate group. They are represented today only by the maidenhair tree (*Ginkgo biloba*), one of the obligatory 'living fossils' to be found in every botanical garden, and many parks in inner cities. Ginkgo leaves are particularly beautiful, perfect fans with splayed veins, and specimens of this general type are found in many localities from rocks of Triassic age onwards. The Mesozoic ginkgos were a varied and numerous group of plants, but unlike the spruces and pines they slowly declined in variety throughout the Tertiary.

BELOW: Varied shapes of leaves of fossil ginkgos, typically Mesozoic. These leaves represent the extinct relative of the living maidenhair tree, *Ginkgo biloba*. The specimen originates from the Middle Jurassic, North Yorkshire, England.

Flowering plants – Division Anthophyta

The flowering plants dominate the floral world today, except perhaps in the lonely spruce stands of the Arctic. They are the broad-leaved trees, the grasses and myriads of tiny creeping plants, as well as the showy flowers which attract humans and pollinating insects alike. Their diversity is truly extraordinary: in tropical forests there are hundreds of species of trees alone. Yet a good deal of this diversification probably happened in the last 65 million years, during the Tertiary. However, like the mammals and the birds, the flowering plants certainly originated, and probably did much of their evolutionary groundwork, during the Mesozoic and especially the Cretaceous. The details of their origins are still not clear. For one thing, flowers do not readily fossilise – and leaves alone can be misleading. The flowering plants (angiosperms) appear almost ready-made in the Cretaceous, and many of these fossils can even be placed into living genera. There is still much to discover of their even earlier history, and we hope that the crucial steps did not all take place in some site where fossils have little chance of preservation.

OPPOSITE: A fossil poplar leaf, *Populus latior*, from the Miocene, preserved in limestone laid down under freshwater. This species has a finely-toothed margin; a large wide leaf borne on a long stem. Length is 18 cm (7 in).

Whatever their early history, fossils of flowering plants – leaves, seeds and wood – are common in rocks of Tertiary age. They do not even have to be rocks deposited under freshwater conditions, because wood and seeds are perfectly capable of drifting long distances before becoming waterlogged enough to sink to the bottom of the sea. Leaves, however, tend to be preserved in the deposits laid down in lakes and the like. Where the sediments are fine-grained the leaves form fossils of exquisite beauty, preserving even the finest veins, which can be directly compared with the leaves of their living relatives. Another mode of preservation of Tertiary plants is beneath flows of lava (Oregon and the island of Skye, for example), where whole floras can be preserved in enough detail to reconstruct the detailed botanical ecology of the time. Fossil plants are of great use in plotting the many shifts of climate that happened during the Tertiary: a humid-tropical flora may be found in what is now an arid region, or a warm-climate flora may be found in an area that is now decidedly cool. This makes the assumption, of course, that the plants have not changed their habits since they were fossilised. During the dramatic climatic fluctuations of the last ice age – warm to cold to warm repeated several times – the flowering plants acted as thermometers for the climate, sensitive recorders of the shifts that affected everything from beetles to humans.

ABOVE: A fossil flower, *Porana oeningensis*, Miocene. These beautifully preserved flowers are from the Oeningen deposits, Germany. Flowers are rare as fossils because they are extremely shortlived. Diameter is about 2 cm (¾ in).

Reviving fossils

In this chapter some of the ways palaeontologists determine how fossil animals lived in the past are described - the methods of reanimating the dead fragments to build up a living creature. Since popular ideas of life in the past are often founded on coloured reconstructions of 'The world in the Jurassic' and the like, it is important to remember that these imaginative scenes are all inferences from bones and other fragments.

RIGHT: The discovery of 'feathered dinosaurs' in China gave us a whole new interpretation of what dinosaurs may have looked like. This juvenile dromaeosaur was covered in branched feathers.

In fact there is nothing fixed about such interpretations, because the way the fossils are understood may change over the years, and it is usually some time before new discoveries percolate into popular presentations. Even the most solid-looking dinosaurs may have changed their habits in the last few years!

Many different lines of evidence may be used to flesh out the bare bones of the fossils. The first of these is evidence from the rocks from which the remains were recovered. The sediments themselves reveal much about the environment of deposition, as was shown in Chapter 3. It is important to establish whether the fossil animal actually lived in the environment that furnished its sedimentary cover, or whether its remains were swept in from some other place. Fortunately it is usually easy to spot such intruders. The type of sediment and the associated fossils show whether the environment was marine, freshwater or terrestrial, providing the basic information into which the ecology of the animal has to be accommodated. The sediments themselves may have preserved some of the tracks left by the animal to give direct evidence of its past activities. From the tracks alone it is possible to be certain of the bipedal stance of certain dinosaurs, and to measure their stride. The character of the rocks, and their setting at the time when they accumulated, provide evidence for the climatic regime in which the extinct fauna lived. Climate imposes certain restraints on possible modes of life; savannah animals differ from those of tropical rain forests, and these again from inhabitants of the tundra at high latitudes.

RECONSTRUCTING FOSSILS

Looking at the fossil animal itself, the first necessity is to reconstruct it as accurately as possible from its fragmentary remains. Sometimes this is a very complex business, particularly for vertebrates with large numbers of small bones. To proceed from the reconstruction to an assessment of probable life habits, two different but complementary approaches are used. One method attempts to compare the structures of teeth or limbs or some other feature of the extinct animal with living analogues. These do not have to be biologically related organisms, the basic argument being that structures that are similar were probably adapted to a similar function. Sometimes these structures are obvious: the ferocious teeth of a predatory dinosaur are a sure indication of hunting habits, with their edges honed into cutting instruments. The structure and function of grinding or chewing teeth in mammals, in which opposing teeth cooperate in action, and which can be matched in extinct, unrelated mammals, is a much more subtle matter, involving detailed studies on the operation of living dental systems to help elucidate the functioning of fossil ones.

The technique of 'hunt the analogue' is a favourite one practised by palaeontologists, but is certainly not foolproof, because there are many fossil animals that defy comparison with living organisms, and some analogues do not stand up to detailed scrutiny. A second method tries to analyse the structure of the fossil. If the fossil is constructed in a certain way, then there are only a limited number of jobs that the structure could perform. The idea here is to decide which of the possibilities is the most likely. This assumes that nature only manufactures efficient designs, and for the most part this is a reasonable assumption. Most animals today do seem to have bodies that accord well with the functions they have to carry out to survive: flyers are aerodynamically efficient, active swimmers have suitable streamlining, herbivorous mammals have teeth appropriate for grinding plant food. Ideally one could construct a model of the fossil to test out these various functions, but the number of examples where the analysis has been pursued this far are limited. Of course it can never be known whether the right answer has been reached (unless somebody dredges up a 'living fossil'); there are only varying degrees of probability. The best answers are obtained where the functional design of the fossil points to the same life habits as a living analogue, and where both are consistent with the geological circumstances in which the fossil remains are found.

SWIMMING TRILOBITES

There are thousands of different kinds of trilobites. All of them are marine and all of them are extinct. Since many kinds of trilobites coexisted at any one time they were probably occupying different ecological niches, behaving in different ways, corresponding with the wide variation in shape that they show. We can use the methods described above to elucidate some of these occupations, and get a glimpse, albeit an imperfect one, of the trilobite as it lived. The following example is one where both methods can be used, and where all the evidence points in the same direction, so the answer is probably correct.

Most trilobites are oval shaped (see p. 109), being rather longer than wide, not greatly convex, and with eyes that occupy perhaps a quarter of the length of the head. There are a few Ordovician trilobite species, however, with enormous, globular eyes. By examining the way these animals are put together it is possible to suggest a likely mode of life.

We can take the eyes first. The trilobite eye is a compound type, and each lens is made up of a calcite crystal. From the optical properties of this mineral we know that the lens is able to interpret light coming from a direction more or less at right angles to the lens surface. We can therefore deduce what the field of vision of our giant-eyed Ordovician trilobite was, by looking at the directions in which all the lenses face. It turns out that our animal was able to see in almost every direction – upwards, downwards, sideways, forwards, and even

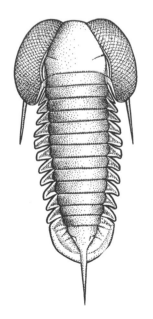
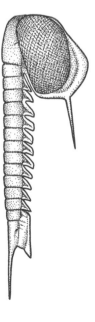

LEFT: Giant-eyed Ordovician trilobite *Opipeuterella*, seen from the top and side (2½ times natural size).

FIELD OF VISION

backwards, because the eyes bulge out beyond the line of the rest of the body. Most other trilobites have a predominantly lateral field of view. If the animal lived on the sea bottom, it seems unlikely that it would have had eye lenses specialised for looking downwards, and so we begin to suspect that the animal habitually dwelt above the sea floor. Other features of its shape are consistent with this. At the edge of the eyes is a pair of long spines, and these point downwards at a steep angle, at a very awkward attitude for resting comfortably on the sea floor. Most 'normal' trilobites have a more or less horizontal rim around the forward margin, which may have rested on the sediment surface.

The thorax of the giant-eyed trilobite is remarkably long compared with the average trilobite. The convex, middle part of the thorax contains the musculature that operates the appendages (which, as usual, are not preserved), and the relatively large volume of this region shows that the musculature was powerful. The construction of this giant-eyed trilobite suggests that it lived an active swimming life well above the sea floor, and possibly in the surface waters of the sea. Because there are no living trilobites we cannot find direct confirmation of this hypothesis, but we can look for other arthropods in the present-day oceans that have the same modifications of the eyes. The analogy is found in the deep-sea crustaceans *Cystisoma*, which also have enormously expanded eyes, looking like headlamps, compared with their bottom-dwelling relatives. In fact many of the arthropods that inhabit the water column have large, globular eyes of this kind. If it is correct that our trilobite lived above the sea bottom, actively swimming in the water, there are certain predictions we can make about its geological occurrence that can be tested by looking in the rocks. Most bottom-dwelling trilobites preferred to live at a particular water depth, or on a particular type of sea bottom (mud, sand or lime). No such restriction should apply to our globular-eyed species; it should be found along with all other different kinds of trilobite assemblages without preference. This has been proved in several places: in some Arctic localities it is found with trilobites that lived at great depths in the muds of the Ordovician ocean, while in Canada the same species occurs mixed with the inhabitants of the shallow water seas, where limestones were accumulating. By the same token the free-swimming trilobite may be expected to have a very wide geographical distribution, for oceans would provide no barrier. Again this proves to be the case: our example is found in Arctic Canada, in the Arctic island of Spitsbergen, in the deserts of Nevada and Utah in the USA, in western Ireland, in Russia, and in northwest Australia.

All the lines of evidence described above suggest that these giant-eyed trilobites of the Ordovician were active swimmers in the surface waters of the oceans including: detailed consideration of the way the trilobite is constructed, analogy with living animals with similar adaptations, evidence from the rocks and the distribution of the fossils.

There are a number of questions we can ask about the life habits of these trilobites which are not subject to such careful scrutiny, for example what did they eat? It is not possible to see directly how the mouth appendages functioned in feeding, and the stomach contents are not preserved. So here we can only speculate. Active swimmers in the surface waters of the present oceans are likely to feed directly on plankton, and the trilobites may have had a method of harvesting large quantities of such food. Alternatively, they may have been hunters of larger prey, in which case when the appendages are eventually discovered they may prove to have adaptations for grasping and manipulating larger food. Puzzles remain, even though we can be confident of the rudiments of the story.

CONTROVERSY ABOUT THE GIANT DINOSAURS

The giant sauropod dinosaurs like *Brachiosaurus* were the largest land animals the world has seen. They weighed more than 80 tonnes and present quite different problems in the interpretation of their life habits from the diminutive trilobites. Such spectacular animals have obviously attracted much attention, and one might expect that the problem of how they lived would have been satisfactorily solved long ago. Many old-fashioned popular books showing the Mesozoic giants in their natural setting portrayed them wallowing about in swamps flanked by deep vegetation, their bodies largely under water. Surely an animal of this bulk, it was argued then, must be partly supported by water, and their relatively inconsequential teeth must have been adapted for chewing on the kind of soft, luxuriant vegetation that flourishes in and around swamps. Because their nasal openings were on the top of their heads, they could even continue to breathe if it became necessary to submerge themselves totally.

More recently the life habits of these giants have been looked at in a way that disproves most of these erstwhile notions. Consider the structure of their legs. The sauropods have relatively long, pillar-like legs, resembling those of the elephant, the largest living land animal, and may have been well adapted for supporting the huge bulk of the animal. The feet of the sauropod are small (relatively speaking), with short, stubby toes, yet animals that walk on soft mud tend to have spreading feet to distribute their weight more evenly. It is difficult to see how the compact feet of the sauropod could avoid becoming stuck fast in the soft muddy bottom of a swamp. If the dinosaur did, after all, live on dry land, then the long neck could have usefully functioned to allow the animal to browse the high foliage of trees. This is how

BELOW: One set of dinosaur trackways seems to show *Apatosaurus*, a sauropod dinosaur, walking on its front feet only! Scientists concluded that it was floating in water, pushing itself along with its front feet and steering with its back legs.

RIGHT: Footprints and trackways tell us whether dinosaurs lived alone or in groups, and how fast they moved.

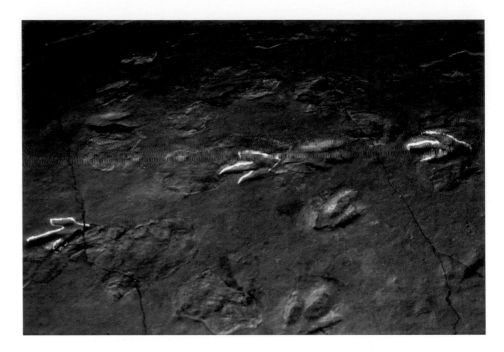

RIGHT: Footprints and trackways tell us whether dinosaurs lived alone or in groups, and how fast they moved.

the animals are portrayed in recent films such as *Jurassic Park* and *Walking with Dinosaurs*. Their fossil remains occur with other animals and plants that are generally accepted as being terrestrial. There is some evidence from the tracks they have left behind that *Diplodocus* and its allies moved about in herds. They may have been the gigantic reptilian analogue of the elephant, and it may be no coincidence that the elephant also has its nasal openings on top of the skull, with the nostrils in this case sited at the end of the trunk – it has even been suggested that some sauropods may have had a proboscis of some sort.

The balance of evidence overwhelmingly supports fully terrestrial habits for these giants. Other aspects of the dinosaur living habits have been the subject of debate. There was much controversy as to whether the dinosaurs as a whole were cold-blooded, like all living reptiles (and there is no doubt that dinosaurs were reptiles), or warm-blooded, resembling mammals. Cold-blooded animals have to 'warm up' before they can be fully active; that is why lizards and snakes bask in the sun in temperate climates. For this reason they cannot cope with climates having greatly extended winters. Warm-blooded animals have the same body temperature at all times, and can be continuously active, but they use far more energy and hence need more food than cold-blooded animals of the same size. The posture of many dinosaurs, particularly the carnivorous theropod dinosaurs, was fully erect with the legs beneath the body, unlike the sprawling legs of living reptiles. The long back legs of such hunters look highly suitable for running, and as they did so the long tail may have been held erect as a kind of counterbalance. For any kind of prolonged activity, warm-bloodedness would have been a distinct advantage. There was, until recently, a vigorous argument between the cold-blooded and warm-blooded schools. Detailed study of the structure of the bones of dinosaurs has found favour in both schools of thought. When they were young, it is likely that dinosaurs grew very quickly, and their bone structure is consistent with a kind of warm-bloodedness like that of mammals today. However, as they grew large, growth slowed down dramatically, and their huge size allowed for a considerable degree of temperature control. They managed to have the best of both worlds, combining reptile energy efficiency with mammalian levels of activity.

The warm-blooded school allied themselves at one time with the theory that the dinosaurs included the ancestor of living birds. The Jurassic bird *Archaeopteryx* is one of the most famous fossils, and was a contemporary of the dinosaurs. Some of the smaller, highly active dinosaurs were about the size of a chicken, and there is more than a passing similarity between a running ostrich and the kind of reconstruction that shows fleet-footed,

BELOW: Four-legged animals with a backbone walk in one of three ways. Monitor lizard (left): sprawling, with legs at right angles to the body; the early ancestors of dinosaurs walked like this. *Euparkeria* (centre): legs straighter and body held high off the ground; this can only be maintained over short distances. *Triceratops* (right): straight legs tucked under the body; the upright stance was the key to the dinosaurs' success.

running dinosaurs. The bird–dinosaur connection got a tremendous boost at the end of the 20th century with the discovery of a whole series of 'feathered dinosaurs', which proved a connection that anatomists had previously inferred from the study of bones alone. Feathers are, of course, otherwise the exclusive property of birds. These important fossils were discovered in China, and there has been a race to describe the most spectacular example – one was even exposed as a fake. Few scientists today question the bird–dinosaur connection. Most specialists regard the dinosaurs and the early birds as having similar specialised metabolism. Truly warm-blooded birds may have arisen as they mastered the difficulties of the aerial habitat. Possibly they retained the same high metabolic rate as has been ascribed to young dinosaurs.

GRAPTOLITES – FLOATING COLONIES

Around 1830, geologists were beginning to unravel the mysteries of the early Palaeozoic rocks. The past world recorded in the rocks could not have seemed more alien. Many of the organisms that left fossil remains were now extinct, although some, like the trilobites, could obviously be placed into a phylum with many living representatives. The graptolites, however, were initially completely enigmatic; indeed they were first described as plants. Their remains were so abundant that they could not be ignored. Their colonies, looking like miniature hacksaw blades, often completely covered bedding planes, and were usually found in the absence of other kinds of fossils. It became apparent that they could be useful in subdividing the intractable stretch of time from the Late Cambrian to the Silurian as they changed in obvious ways from one rock formation to the next. Graptolites with numerous branches seemed to dominate the earlier rocks, ones with fewer branches appeared later, while in rocks we would now recognise as Silurian and Early Devonian forms with a single branch (or stipe) were abundant. With the discovery of better-preserved material it became apparent that the graptolites consisted of rows of tiny cups which were interconnected by a common canal and that they were colonial animals. As described above, we now know that the graptolites are an extinct branch of the phylum Hemichordata, an insignificant group today, consisting of a few encrusting colonial organisms.

How did these mysterious organisms live? In this case it is difficult to apply the analogy method, because the graptolites are different from any animals now alive. The modern

OPPOSITE: Two 'feathered dinosaurs' *Confuciusornis sanctus*, 124 million years old from Liaoning Province, People's Republic of China. Thousands of individuals recovered from this site suggest that they might have lived in large colonies.

BELOW: Growth and budding of an Ordovician graptolite colony, enlarged about 20 times. They grow from a single tube or sicula, by budding.

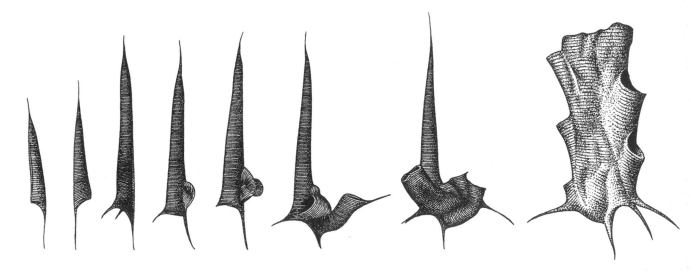

PALAEONTOLOGICAL ENIGMAS

Every now and then the fossil record throws up fossils which are palaeontological puzzles. They are obviously the remains of some kind of animal, but the problem is to decide what kind. They tend to be rather rare and preserved in a special way. Like so many palaeontological matters, they stir up arguments between specialists who think they have a way of solving the enigmas.

Some of these puzzling fossils are quite small. A few years ago a minute fossil only about 2 millimetres (less than ¹/₁₀th inch) long was recovered from limestones of Ordovician age, and christened *Janospira*. It looks remarkably like a trumpet. The 'mouthpiece' and the 'horn' of the trumpet are both open, and a coil hangs down from the middle. The end of the coil is closed, and it seems reasonable to assume that the animal started growing as a coiled shell. It must have then changed its mind, and started to grow in two directions, a narrower tube into the mouthpiece and a broader one into the horn, and there is some evidence that the horn end continued to get wider and longer. The problem is that it is hard to equate this kind of growth with that of any known animal group. It looks like some kind of mollusc, but no mollusc fits easily into this pattern of growth; some have suggested it might be some kind of snail, or perhaps a monoplacophoran. It remains a puzzle. It is found along with fossils of planktonic organisms, and it is possible to explain the change in growth between coil and trumpet as a change that happened when the larval shell settled on the sea bottom. But since virtually all phyla of animals are found with planktonic larvae, this is no help in solving the puzzle.

Some larger fossils are even more puzzling. A number of fossils from the famous Mid-Cambrian Burgess Shale of western Canada are so strange that they have even been claimed as the representatives of 'extinct phyla'. One of these, *Hallucigenia*, is an animal with what is thought to be a gut with a set of tubules arising from it, and below this are curious pointed structures. A suggestion was made that early on in metazoan evolution there were a number of experimental designs that did not give rise to direct descendants; naturally these do not fit into the 'pigeon holes' based on living animals. *Hallucigenia* was claimed as one of these. However, subsequent studies of this oddity have shown that the early interpretation was completely wrong. It was upside down! Further investigation revealed that the tubes on the 'back' were actually legs, and the pointed structures were spikes on the back of the animal. *Hallucigenia* is related (distantly) to the velvet worm *Peripatus*. It is still a very odd creature, but its place in nature is now explicable. One of the fascinating things about palaeontology is that new explanations of such puzzles cast new light on the history of life. Another long-standing palaeontological enigma has only been solved since the third

ABOVE: *Hallucigenia sparsa* from the Mid Cambrian Burgess Shale of western Canada.

BELOW: Older (top) and more recent (bottom) reconstructions of *Hallucigenia* turning it upside down.

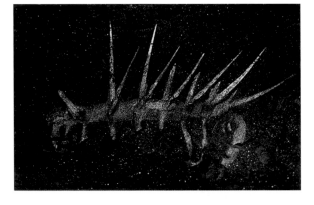

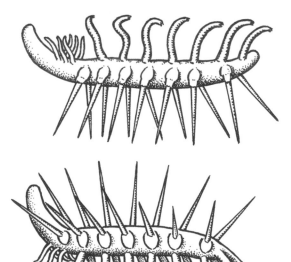

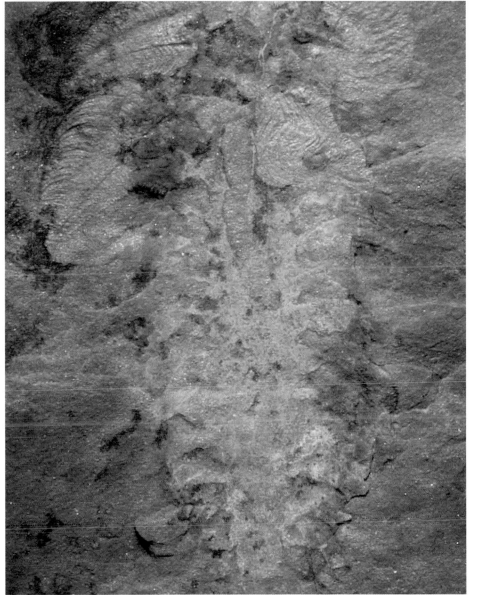

LEFT: An Ordovician machaeridian, an armoured invertebrate, *Plumulites bengtsoni*, from Morocco was recognised in 2007 as an annelid 'worm' for the first time.

edition of this particular book has been published. For more than a century Ordovician and Silurian specialists had puzzled over small, triangular calcareous plates with a rather delicate feathery ornament. On occasion these plates were found arranged in lines, as if they were 'tiles' or covering up some larger animal; they were called machaeridians. Learned professors argued over what they might be. Some drew a comparison with a group of primitive molluscs called chitons which have a covering of plates; others thought of echinoderms with their tests of many plates. Nobody really knew the answer: that is, until in 2007 a specimen with legs was found in Morocco, which proved that the machaeridians were actually annelid worms – of an armoured variety without any ready living analogue. There is something rather wonderful about fieldwork still yielding finds that no amount of armchair speculation could have anticipated.

Origin of life

When the Earth formed some 4,550 million years ago there was no life. By about 3,500 million years ago the first, tentative traces of life are to be found in the rocks. The profound series of changes that have occurred since the Cambrian, and which are the subject of most of this book, took only a fraction of time in relation to the history of the Earth.

RIGHT: These stromatolites are formed from blue-green algae. Over the last 4,000 years, algae have trapped detritus and sediment creating large microbial mats. The algae secrete calcium carbonate causing the mats to mineralise, forming the rock-like structures. Hamelin Pool Marine Nature Reserve, Western Australia.

The time of the origin of life is very inaccessible to direct study, all certainties about the nature of the atmosphere and configuration of land and sea are gone, and speculation has more room for manoeuvre with this palaeontological mystery than with almost any other. What is certain is that the Earth's surface, when life originated, must have been different in almost every respect from that of the present.

THE EARTH AND ITS ATMOSPHERE

The formation of the Earth was but one incident in the formation of the solar system as a whole. If present ideas are correct, the Earth accreted from small particles swirling in a disc-shaped nebular cloud. The planet grew from a 'seed' of magnetic metallic grains, which attracted more material by gravity, so that it grew rather like a snowball. Early on, the Earth melted and a molten core was formed, largely composed of the metals iron and nickel. Accretion of other material continued, with the addition of volatile components like water, carbon dioxide and chlorine when the body was large enough to retain them. Major bombardment phases ensured that the growing planet was continually sterilised and was prevented from retaining much of an atmosphere at this time. Then the surface of the Earth cooled quickly to form a thin, brittle crust. But massive bombardment with meteorites

RIGHT: The western hemisphere of the Earth's surface in 1997, created from the data of three different Earth observing satellite instruments. At the time of the origin of life the Earth's surface would have been completely different to what it is today.

continued; the surface of the moon preserves this stage in the evolution of the Earth as it is covered with impact craters. The moon itself may have been the result of a particularly huge impact on the 'proto Earth'. Our planet incorporated the additional meteoritic material into the upper part of its mass. The gravity of the Earth was powerful enough to hold on to some of the more volatile ingredients, particularly water, without which there would have been no life. An atmosphere gradually formed around the Earth, and the incessant rain of meteorites slowly abated. The rate of heat flow from the Earth's interior was high at this early stage, and volcanic activity was widespread and continuous. Any volcanic eruption is accompanied by the release of huge quantities of steam and gases like nitrogen and carbon monoxide, as well as highly toxic acids like hydrogen chloride and sulphur dioxide. We can visualise acid rains attacking rocks and reacting with them to form minerals such as sea salt (sodium chloride). The steam released into the atmosphere from the volcanoes condensed to form seas, and since the first traces of sedimentary rocks are as old as 3,800 million years or more, there was enough water for sedimentation by this time. Presumably the processes of erosion had also begun to shape out patterns of cliffs and beaches so that the first vestiges of the modern style of topography appeared. Even so, it was an alien world.

In several important respects the environment of the early Earth was completely different from today's environment. The original atmosphere may have had a much larger proportion of carbon dioxide than is present in the current atmosphere. There was little or no free oxygen in the atmosphere. This meant that there was no ozone layer (a cloak of modified oxygen high in the atmosphere today that acts as a screen to prevent harmful radiation getting through to the Earth's surface). The penetration of such light to the surface of the Earth produced the kind of chemical reactions that might have led to the necessary building blocks for life itself, including the splitting of the water molecule into its component atoms. These chemical reactions can be reproduced in the laboratory – the right mixture of chemicals and gases, charged with the effects of ultraviolet light – and sure enough some of the basic ingredients of life appear after a short while. Nonetheless, the actual moment of creation of the first self-replicating entity that might be called living is not recorded in the rocks. We are driven to speculation.

EARLY LIFE

The presence of amino acids in the early Earth was particularly important, for chains of such acid molecules join together in their hundreds to produce giant molecules, forming the basic structure of proteins, without which there would be no life. All organisms are able to produce copies of themselves, and presumably it was necessary to have evolved the long spiral chain molecule DNA (or possibly RNA first), which is the basis of the copying process, before the first self-replicating cells could exist. The first living cells were of the simplest kind, represented today by bacteria (especially Archaea). Many primitive living bacteria have an 'economy' built on exploiting sulphur compounds, which were present in abundance around the hot springs, fumaroles and volcanoes in the Precambrian. They are tolerant of high temperatures – in fact they need heat to survive. Studies on the genes of such organisms show that they are likely to be the most primitive organisms alive on Earth today. Life probably started at high temperatures, in the absence of oxygen. Whether the vital steps occurred in the early seas or in some other location, such as shallow pools or the hot vents of submarine volcanoes, is still a matter of debate. But presumably once the first true, self-replicating cell had appeared it could spread and prosper unhindered wherever the right conditions for its nourishment were to be found. That Earth, no matter how inhospitable it would seem to us, would have been a Garden of Eden for the first organism. Most living things have so many molecular features in common that it seems feasible that the generation of life happened just once. We are only now beginning to appreciate the huge variety among the bacteria. In some respects they are the most successful organisms on Earth, for they are found everywhere from the Arctic to deep beneath the surface of the Earth in the rocks themselves.

The origin of life requires principally time, the right conditions, and a few special events to link all the components into the finished cell. Recently another idea has claimed a lot of attention, although it is not a new one. Life may have been of extraterrestrial origin. The early Earth was almost certainly bombarded with particular kinds of meteorites (known as carbonaceous chondrites) and these meteorites when they appear at the present time do contain organic kinds of molecules. As far as the origin of life on Earth is concerned it is true that it is possible to produce carbon compounds like those in meteorites under the right experimental conditions, but there is a curious difference between these compounds and those that predominate on Earth. Many carbon compounds can exist in two forms, which are chemically identical, but which have the property of rotating light either to the left or to the right. On Earth there is a predominance of the former that rotate light to the left, but in meteorites the left and right-handed forms of the same compounds are equally abundant. It

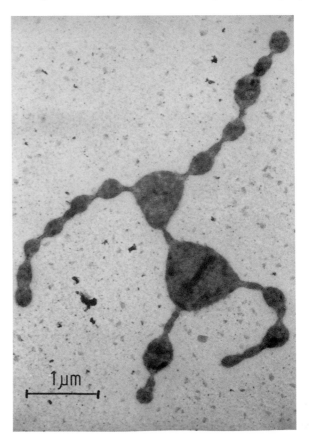

RIGHT: Hyperthermophilic Archaea which lives in hydrothermal vents at temperatures around 100°C (212°F).

therefore appears as if organisms on Earth manufacture carbon compounds that are subtly different from those produced in meteorites. It seems preferable to regard the Earth as the cradle of life, if only because this theory requires fewer coincidences to produce an organism adapted to terrestrial conditions, and there has been a vast stretch of time available for the shuffling and rearrangements of chemicals necessary to make the first living cell. Therefore, although meteorites contributed organic molecules to the 'soup' from which life was brewed, it is probable that the cauldron itself was on the Earth and not in the far reaches of space.

Some scientists have claimed that the periodic appearance of extraterrestrial materials on Earth, especially from comets, has continually influenced the evolution of life – not just its earliest history. For example, they suggest extraterrestrial biological material provides an explanation for catastrophic viral epidemics. Such meteorite-induced epidemics are one of a number of theories about extinction that it is impossible to test one way or the other. It is difficult to see how meteorites could have been so selective as to remove, say, dinosaurs but not birds (which may be close relatives of the dinosaurs), or ammonites but not snails at the end of the Cretaceous. Comets passing close to the Earth from time to time may well be implicated in some of the crises in the history of life, but this is more likely to have been by secondary effects rather than by direct biological influence. Viruses are rather specialised organisms, often adapted to a particular host, so a virus with fatal properties arriving from 'out there' might be expected to remove only its host species. This does not explain mass extinctions as we actually observe them in the rocks.

One way or another the crucial steps were taken, and cells evolved. It is only in the last few decades that proof of the existence of the oldest kinds of organisms has been recovered from the rocks. As one might expect these rocks are rather special: they have survived from the remotest parts of the Precambrian hardly altered by heat or pressure. They are often cherts, fine-grained hard rocks composed of silica, and when cut into thin slices and ground they become transparent and reveal tiny fossils. Most of the cherts that have yielded fossils seem to have been deposited in shallow water around the proto-continents of the time.

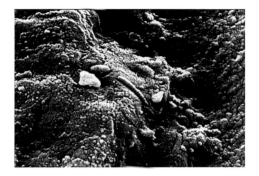

ABOVE: SEM image of the surface of the carbonate patches in the Martian meteorite ALH 84001, showing a tube-like 'microfossil'. The organic nature of these microfossils is still disputed.

PRECAMBRIAN ORGANISMS

The earliest and most primitive fossils are those of prokaryotes. These are simple rods, spheres or filaments in which the cell contents lack a defined nucleus (the package within the cell which houses the DNA in all higher organisms). Prokaryotes would be referred to as bacteria today. Such simple prokaryotes are known as fossils in rocks as old as 3,500 million years. Some of the earliest fossils are so simple that they still excite controversy among scientists as to whether they are really of organic origin. However, uncontroversial fossils of eukaryotes (those organisms with a defined nucleus) are known from rocks about 2,500 million years old. Even for those used to thinking in terms of geological time these figures are hard to comprehend. The time during which these simplest of organisms held sway is so much longer than that which has elapsed since the beginning of the Cambrian, which used to be thought of as comprising the whole of the fossil record. Momentous events at the cellular level happened during this vast time. The earliest fossils discovered from cherts can scarcely represent the whole story. New discoveries are still being made, and it is clear

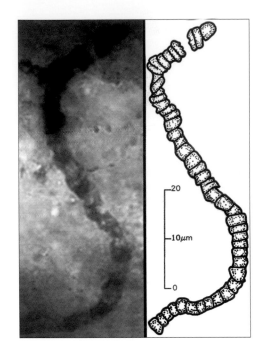

that there was an increase in the variety of these small fossils through the Precambrian, even from the patchy record they have left.

Ancient bacteria were able to live in conditions which would be inimical to most living organisms. Some had a metabolism based on sulphur, and these survive today in hot springs and similar habitats. Others acquired the ability to use the sun's energy to produce nutrients (the process of photosynthesis), but not all of these gave off oxygen, as do the higher plants. Cyanobacteria that did exhale oxygen were certainly present 2,700 million years ago, and some scientists claim that they may even have been present 3,500 million years ago.

Some Precambrian fossils had been known about for well over a century before it was concluded that they really were organic. These include finely layered rocks, often with the external appearance of many small cushions, or occasionally steeper pillars. The regularity of their layered structure first suggested they might be of organic origin: such fossils were christened with names like Cryptozoon ('hidden animal') when they were discovered in Precambrian rocks from Canada and elsewhere. At the time it was considered rather unlikely that fossils could occur in rocks as old as this. Nowadays, fossils of this kind are known to be the remains of structures produced by cyanobacteria, a perfectly reliable indication of biological activity. They are known as stromatolites. Structures resembling stromatolites are now known from rocks younger than Precambrian age as well. They are common in limestone rocks originally deposited in shallow water environments in the tropics, and they appear to be especially characteristic of quiet water sites between high and low tides in areas of high salinity. A number of years ago living stromatolite mounds were discovered in Shark Bay, Western Australia (p. 159), and like some of their fossil counterparts they were found in an intertidal environment. Since the oldest of the fossil stromatolites were at least 3,000 million years old, these rather unspectacular mounds take all prizes for the most enduring of living fossils. Why they had not been recognised as of organic origin previously is because

ABOVE: Section through fossil stromatolite cushions from the Precambrian rocks of eastern Siberia (half size).

LEFT: Fossilised bacterial/algal stromatolites around 15 million years old, Wadi Kharaza, northern Egypt.

the very primitive organisms that make the mounds are not usually preserved as fossils – their simple threads generally decay without trace. The mounds are produced by a skin of tacky, slime-covered cyanobacteria trapping sediment season after season to produce the fine layering so characteristic of the stromatolite.

Beneath the surface layer other, ancient kinds of bacteria thrive, so that the stromatolite is really a very simple community, one which has survived from the earliest days of the Earth. Tidal waters drain off through the channels between the mounds. In the living examples there is a certain variation in the form of the mounds according to where they are growing. Some of the same variation has been observed in the fossil forms. Nonetheless, a growing number of palaeontologists studying the Precambrian mounds believe that there is also a variation through time in the shapes of the mounds, so that they can be used to characterise very broad segments of the Precambrian. Their occurrence in Precambrian rocks which were deposited over the shallow, continental seas of that time is almost ubiquitous; they have been found over huge areas of Russia, Australia, Greenland, Africa, Canada, the USA and Scotland. They were much more widespread in these far-off times than they are today and some of them occurred in deep water sites. This is probably because there were no Precambrian animals adapted for grazing, one of the standard ecological niches in living marine faunas. Living stromatolites owe their existence to special conditions inhibiting such grazers. It is not altogether fanciful to visualise vast stretches covered with stromatolites, enduring for a period of considerably more than 2,000 million years of the Precambrian.

BELOW: Living algal stromatolites on the sea floor, Exuma Cays, Bahamas.

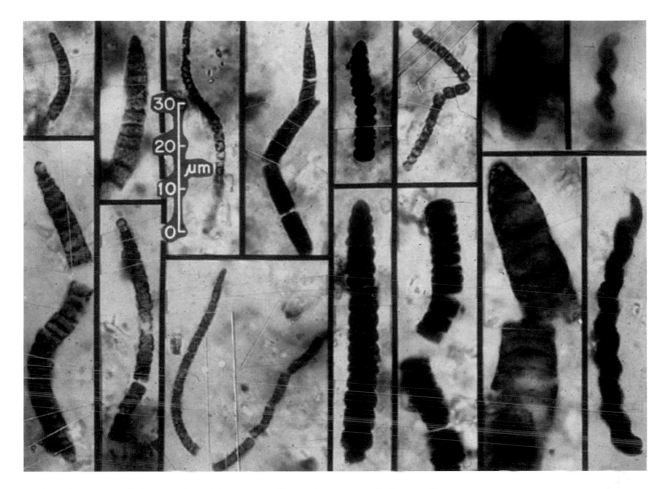

Cyanobacteria and algae produced free oxygen during the process of photosynthesis. This process was probably the most important ingredient for setting up the conditions for the evolution of all higher organisms. The photosynthetic activities of cyanobacteria and algae proceeded for more than 2,000 million years, and if they were as widespread as seems likely they were capable of adding oxygen to the atmosphere. An atmosphere was finally created in which other organisms could breathe; they made the environment fit for animals. It is possible to trace the enrichment of the atmosphere in oxygen by changes in the types of sedimentary rocks that were laid down during the Precambrian. In the earlier part of the Precambrian when relatively low oxygen conditions prevailed, there are widespread deposits of a distinctive kind of banded iron ore. Iron has a great affinity for oxygen (it 'rusts'), meaning that iron scavenged much of the oxygen produced by the early photosynthesisers, which was then taken out of the atmosphere in the ores. Actual increase in atmospheric oxygen was probably held in check for a long time by this process. However, these banded iron ores are almost never found in rocks younger than 1,000 million; the few younger occurrences are small and isolated. There must therefore have been a critical point when there was enough oxygen in the atmosphere to permit respiration; it was possible for animals to exist, and perhaps begin feeding on the plants that had made their existence possible. At the same time the atmosphere would have acquired its protective ozone layer, which would cut out some of the more harmful effects of solar radiation.

ABOVE: Late Precambrian Bitter Springs Chert, about 850 million years old, has yielded a variety of micro-organisms, including the spiral *Heliconema*, with cells apparently in the process of division.

their stomach cavities are occasionally preserved as casts in fine sediment. A few of these jellyfish may even be related to forms living today, and it is reasonable to assume that they had similar habits, drifting about in the open oceans carried by currents, feeding on planktonic organisms. Their stratigraphic position below the base of the Cambrian has been known for some time; *Brooksella*, a jellyfish from the Precambrian rocks at the base of the Grand Canyon sequence, was discovered many years ago, while some of the Canadian examples have been known for almost a century. It was some time, however, before they were accepted as 'real' fossils. Even now, some of the supposed Precambrian jellyfish have been shown to be a rather peculiar kind of sedimentary structure, entirely inorganic, produced by the expulsion of water from wet sediments. However, more and more examples are now being found, many of them indubitably true fossils, and their worldwide distribution is impressive: several localities in Canada, the USA, Australia, Russia, and scattered occurrences in Europe, including Britain. We might anticipate that a free-floating organism like a jellyfish would have a wide geographical spread, but it is surprising to find that so many localities had the right sedimentary conditions for their preservation. It is clear that the Late Precambrian oceans supported a variety of planktonic single-celled algae, many drifting jellyfish, and a variety of small, soft-bodied planktonic animals, perhaps resembling the larvae of living marine organisms, which provided food for the jellyfish.

ABOVE: Fossils from the Late Precambrian Ediacara fauna from South Australia. The enigmatic *Tribrachidium* (top) and the possible annelid worm, *Dickinsonia* (bottom).

RIGHT: The soft-bodied enigmatic animal *Fractofusus*.

There are a number of Proterozoic soft-bodied fossil faunas that include a much wider variety of fossils than simple jellyfish. The most famous of these is probably the Ediacara fauna from southern Australia, where there is exceptional preservation of many quite large soft-bodied organisms. The true affinities of these impressions are still under debate. It is possible that segmented worms (annelids) are represented by such fossils as *Dickinsonia*. Some other curious segmented animals have been suggested as arthropods, but this is far from generally accepted. If it is true that the arthropods were derived from a segmented, worm-like animal, then we might expect to find the common ancestor of all the arthropods in the later part of the Precambrian period. The peculiar animal *Tribrachidium*, with its three curved grooves, has been suggested as an ancestor of the echinoderms, but there is really not much evidence to support this. A common element in the Ediacara fauna, and one which has been found in other parts of the world, including Newfoundland, Russia and Charnwood Forest, England, is a frond-like fossil, *Charnia*, which looks rather like the living sea pens. All of these animals have attracted a lot of controversy about their true affinities. It is no simple matter to relate soft-bodied Precambrian fossils to living animal phyla, and the Ediacara fauna has shown that there seems to have been a distinctive Late Precambrian fauna with a number of curious animals all of its own. This impression is confirmed by the discovery of other faunas in Russia and Newfoundland, which include some forms in common with the Ediacara, but have others unique to them. The Newfoundland fauna occurs in a series of beds exposed at Mistaken Point, a promontory on the southeast of the island. Here, centuries of weathering have etched out the impressions left by the soft-bodied animals, which remain like a picture gallery of the distant past, now washed by the waters of the present Atlantic Ocean. The age of the Ediacara fauna is not much older than the base of the Cambrian, but the Newfoundland faunas may be rather older. The peculiar animals were evidently widespread, although rarely preserved as fossils, and may have held sway over the Late Proterozoic sea between 580 and 560 million years ago.

Apart from the jellyfish, most Precambrian animals were probably bottom dwellers. It has been suggested that they housed symbiotic bacteria in a 'Garden of Ediacara'. Even in rocks where the remains of soft-bodied animals are not preserved you might expect to find the traces left behind by the activity of animals in the sediment – burrows and trails. Oddly enough, trace fossils are relatively rare in Precambrian rocks, even though they are common in rocks of Early Cambrian age from many parts of the world. A number of supposed trails have been found in Precambrian rocks (even in excess of 1,000 million years old), and some of these may be genuine. However, similar looking marks can also be produced by inorganic causes and so whether you interpret them as animal or not depends on how much you want to believe that there were animals around at the time. In some sections of rock, in Arctic Norway for example, it is noticeable that the numbers and variety of trace fossils seem to increase as the boundary with the Cambrian is approached. Once the boundary is passed it is easy to find deep burrows, vertical to the sediment surface. Perhaps it is indeed true that burrowing habits were only acquired by animals of Cambrian and younger age.

CAMBRIAN ORGANISMS

No survey of the early history of life would be complete without taking the story over into the lowest part of the Cambrian period. The old idea that life appeared quite suddenly at this boundary will by now be seen to be entirely wrong; we have traced the increase in variety of

RIGHT: Some of the oldest fossils with hard parts that appear near the base of the Cambrian; few of them can be easily related to living organisms (all a few millimetres, less than ½ in, across).

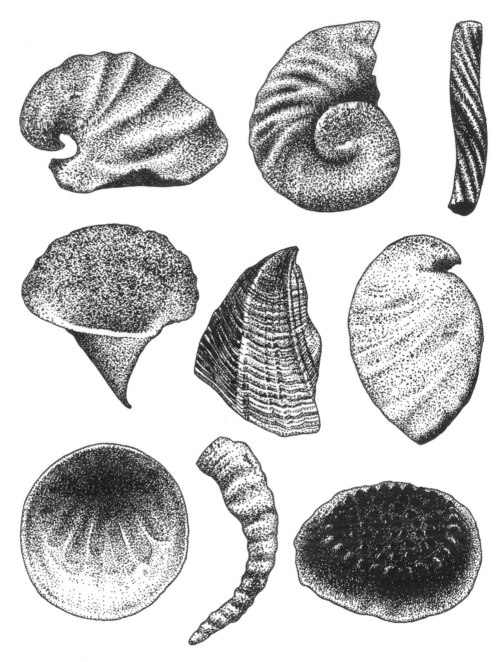

organic remains through the Precambrian from about 3,500 million years ago. The evidence is still tantalisingly incomplete and it is still possible that one discovery which miraculously preserves a whole fauna and flora could change the story in many details. The last 30 years have seen a great increase in the effort put into studying fossils from the earliest Cambrian. Although Cambrian rocks were first discovered in Wales, the earliest beds of the system are not well developed there, and the most important rock sections for studying this interval of time are from other parts of Europe, China, Morocco, Australia, Canada, Greenland, the USA and Siberia. These are mostly sediments deposited in relatively shallow water marine environments. The best approach to understand what happened in the very earliest Cambrian is to collect fossils bed by bed through the critical interval, charting the first appearance of the various kinds of animals the rocks contain.

It is a remarkable fact that animals with hard parts appeared widely at the base of the Cambrian. For several thousands of millions of years there had been life, but no organisms had secreted a shell, or any kind of preservable strut to serve as a skeleton. Within what must have been a comparatively short time period, many different kinds of organisms became able to do just this. It is still something of an enigma that the appearance of these hard remains from so many different sections in different parts of the world was at approximately the same time. What is certain is that it happened, for the majority of organisms, within a very short time period beginning close to 540 million years ago, although somewhat earlier skeletised fossils have been found in Namibia in recent years . Not all of the various kinds of animals that are characteristic of the Cambrian made their appearance at the same time in the earliest rocks. There was a time lapse between the very first hard parts to appear in the rocks, and the appearance of some of the animals that are familiar in later Cambrian and Ordovician rocks. Trilobites are not present among the earliest shells, for example. The organisms that appear first are often peculiar small shelly fossils, many of which are geological enigmas. They simply cannot be fitted conveniently into the groups of animals we know about from younger rocks. A lot of them are simple tubes, sometimes twisted, and with differing cross-sections. They could have been secreted by any number of different worms, or by something else. There are a number of odd, cap-shaped shells, which are even more puzzling. Some of them appear to fit together in a kind of mosaic, and very few of them appear to be obviously related to anything else that follows. Yet some of them are also very widespread, so whatever kind of animals produced them, they were evidently successful for a short time. In some sections these peculiar assemblages of fossils precede the rocks that contain more familiar kinds of animals, like trilobites. Perhaps these early oddities were 'experiments' that had a short heyday, but did not

BELOW LEFT: *Helicoplacus everndeni* from the early Cambrian, Westgard Pass, California, USA.

BELOW: Restoration of *Helicoplacus*; a peculiar and puzzling Cambrian echinoderm.

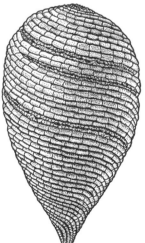

survive long once the typical Cambrian faunas were established. It is also clear that several kinds of small shelly fossils may have belonged to a single, larger animal. The discovery of an entire specimen of the enigmatic animal *Halkieria* from Greenland has shown that many completely different shells were housed on a much larger animal.

Among the Early Cambrian fossils were the peculiar archaeocyathids, a distinctive group of sponge-like forms, which produced the earliest structures that might be called reefs. They made their appearance shortly after the earliest faunas with hard parts, and did not long survive the Early Cambrian. The kinds of limestone that contain archaeocyathids also yield remains of calcareous algae that have secreted a calcium carbonate skeleton around their delicate threads, so it was not only animals that acquired the ability to build skeletons at this time. Trilobites regularly appear among the Early Cambrian faunas, and often there are a number of different species together. The earliest trilobites were already quite complex animals with well-developed eyes. The earliest molluscs are also present in Early Cambrian rocks, and include representatives of the gastropods, chitons and the monoplacophorans, but many of these are hard to trace below the Ordovician. The earliest echinoderms from the Early Cambrian include some very odd animals that are not easy to relate to any of the echinoderms that are prominent in recent oceans, most of which have histories going back to the Ordovician. The novelties include the extraordinary helicoplacoids (p. 173), which, like so many of the more peculiar animals, did not survive past the Early Cambrian. Many of the animal groups which dominate Ordovician and younger rocks have only the sketchiest Cambrian histories – admittedly we can recognise other representatives of the same phylum, but these tend to be rather odd animals, with a history confined to the Cambrian.

There are further peculiarities of these early faunas. A number of Early Cambrian animals, like lingulate brachiopods, seem to have used calcium phosphate as the material to build their skeletons. Among the animals that dominate Ordovician seas right through to the present, calcium carbonate is by far the dominant building material for skeletons (with the important exception of the vertebrates). Two kinds of animals used silica for the construction of their skeletons: some of the sponges and radiolarians. These two groups also have a history extending back into the Cambrian. When skeletons were first produced, it seems as if there were no particular advantages in using one material or another, but subsequently calcium carbonate proved more advantageous, or at least more easily laid down.

So we come to the critical question: why did the appearance of hard parts apparently happen within such a short time? In the past, before so much was known about Precambrian animals, the answer might have been that there *were* no animals until shortly before the Cambrian. The sudden appearance of the various animal phyla would then have naturally been accompanied by the sudden appearance of their shells. Today, there are advocates of such a Cambrian 'evolutionary explosion' who maintain that the faunas of Ediacara type fell victim to a major extinction, and that the Cambrian appearances were the result of a subsequent period of frantic and accelerated evolutionary change. The alternative view emphasises the fact that even in the earliest part of the Cambrian the types of organisms with shells were already highly varied, showing that evolution had been proceeding on a number of independent lines for a long time. It would be hard to believe that such independent lineages should by chance alone 'decide' to place a premium on acquiring hard parts at the same time.

OPPOSITE: One of the earliest trilobites from the Cambrian, already quite a complex animal with well-developed eyes. *Olenellus*, Lower Cambrian, Pennsylvania, USA.

ABOVE: *Helcionella*, a monoplacophoran from the Middle Cambrian of Shropshire, England.

It may be possible to fit the Cambrian events into the broader picture of the evolving atmosphere, and the changes in the Earth as a whole, which were discussed earlier in this chapter. There is some evidence to indicate that it was not possible for calcium carbonate (or perhaps the other skeletal minerals) to be deposited by living tissues until the partial pressure of oxygen in the atmosphere had reached a critical level. Possibly this level was reached near the base of the Cambrian (although many experts place this event much earlier). If conditions suitable for phosphate deposition had been reached at the same time this might explain why there were phosphatic shells also at this period. I have already mentioned that there was also an exceptionally widespread glaciation in the Late Precambrian that spread even into tropical latitudes at about 650 million years ago. It has been suggested that the sudden appearance of ancestors of animals with hard parts may have been related to this event. Glaciation would have lowered sea level, and the cold climate may have produced effects dramatic enough to exterminate the earlier animals lacking shells that dominate the Precambrian. Subsequent elevation of sea level following the melting of the Precambrian ice-caps would have produced a worldwide marine flooding, favouring the evolution of new animal designs, and possibly boosted oxygen levels.

The base of the Cambrian thus marks a change in environmental conditions, and most of the earliest Cambrian animals seem to have been inhabitants of shallow seas that transgressed across the Precambrian landscape. This recolonisation may have come from planktonic animals, or perhaps the deep sea. Planktonic animals and plants, including jellyfish and single-celled algae, would have been best adapted to surviving the peak of the earlier glaciation by retreating to the open ocean, or to unglaciated areas around active volcanoes. If this explanation is correct then the earliest Cambrian would have been a period of extraordinarily rapid evolution – an 'evolutionary burst' like the radiation of mammals in the Early Tertiary. If the ancestors of many of the Early Cambrian forms were planktonic animals, it may not be a coincidence that all major phyla still have planktonic larval stages. The growth of the shell may have been a response to settling on the sea bottom. All this is speculative, and very difficult to test one way or the other. The acquisition of shells and skeletons is one of the great milestones in the history of the biosphere, and the difficulty of finding a single neat explanation only adds to its fascination. Some recent evidence from molecular sequence modelling of a variety of genes exploring the divergence of phyla and classes of animals, does point to a longer Precambrian history of some of the major groups than the fossil record alone indicates. If this is confirmed it means that there are still Precambrian fossils to be discovered which may change our ideas of the early biological history of the planet. On the other hand, discovery of genes that control major aspects of development and design in organisms indicate that small changes in the genes might presage big changes in the organism concerned – and maybe the 'time was right' close to the base of the Cambrian for these genes to be expressed in several different evolutionary lines, producing new designs of organisms in quick succession.

Extinction & evolution

Extinct animals had their day, were somehow unfit for the modern world, and as a result, died out – so the story goes. This chapter explores some of the causes of extinction, and shows that without extinction there would have been little chance for evolution. While the attitude of conservationists to prevent the wanton destruction of species is laudable, the fact remains – in the normal course of events species will become extinct.

RIGHT: Artist's reconstruction of *Diprotodon*, meaning 'two teeth in front'. *Diprotodon* was one of the largest marsupials to have lived. It was hippo-sized and existed during the Pleistocene period.

The fauna of the present day is the product of repeated changes in the faunas of the past, and there is no reason to suppose that the process will stop now that *Homo sapiens* covers much of the world, although there is equally no reason why humans should speed up the process by wholesale destruction of habitats or excessive hunting. As the world has changed, so have the faunas and floras, with evolution and extinction playing their complementary parts in constantly reshaping the biosphere.

There are two types of extinction. In one, the process is as final as in the case of the dodo. A species dies out completely, leaving no progeny. If a whole group of related species become extinct at about the same time, a major animal group may become extinct. There are no dinosaurs lurking in remote corners of the world, not even in Loch Ness. More than anything else it is the disappearance of such major groups that has changed the appearance of the fauna through geological time. A second type of extinction involves the generation of new species. One species gives rise to another, by any one of a number of processes. Ultimately the parent species may become extinct, but in a sense its genetic material lives on in the daughter species. The only totally dead groups are the side branches of the evolutionary tree, for which the long chain that connects the living species with the ancestors from which they sprang has been irrevocably severed.

EXTINCTIONS AND THE FOSSIL RECORD

If extinct species, which are the ancestors in distant geological periods of living fauna, have hard parts and are readily fossilised, it should in theory be possible to read the past simply by collecting fossils through the strata from the youngest rocks to the oldest. Anybody who tried to unravel the evolutionary story directly from the rocks would be disappointed. As Darwin was well aware, the rocks seem to have many gaps and holes in the record. The ancestor hardly ever seems to be sitting there, where it should, in the rocks immediately below the

RIGHT: Skeleton of *Raphus cucullatus*, the dodo, collected from the Mascarene Islands, which include Mauritius, Indian Ocean.

descendent species. Sometimes a species thought to be close to the ancestor of a whole group of animals turns up in surprisingly young strata. The fossil record is imperfect and capricious. This has led a few palaeontologists to reject the stratigraphic order in which fossil species occur as being of no significance at all in the elucidation of the evolutionary process. It is certainly not surprising to find primitive-looking animals in rocks younger than we might expect; after all, the 'living fossils', which are important in unravelling the relationships between different animal and plant groups, are all in a sense organisms that have out-lived their time. Some primitive animals are well adapted to their own ecological niches from which they have never been displaced. Evolutionary events do not happen in a regular, step-wise fashion; it is a rather messy, irregular process, with survivors persisting alongside novelties. Momentous evolutionary steps are often taken by humble organisms, while the dominant animals may undergo spectacular evolutionary extravagances that are doomed to total extinction. Some theorists have even said that almost the whole pattern of evolution may be accounted for by chance alone.

Why does the fossil record seem to be so imperfect? Part of the reason is that there are many breaks in even the most continuous-looking rock succession. Bedding planes are the record of such breaks. In some successions deposition occurs irregularly, such as after a major storm, and estimates suggest the rock may represent as little as 1% of the total time in some sedimentary sites. The evolutionary process is relatively slow, and one might expect to find the evolutionary story visible even through small pieces of the record, just as you can recognise the subject of a pointillist painting by standing at a distance. A more important reason for imperfections in the fossil record is that most rock successions record shifts in facies: the sedimentary (and biological)

LEFT: A model of the dodo formerly from Mauritius – the emblematic animal for extinction.

environment changes with time, and as this happens the evolving animals are carried elsewhere and the descendent species are recovered from a different rock succession, maybe many kilometres away.

Perhaps the most significant factor affecting the fossil record stems from the way new species are derived. Where the generation of new species has been studied among living animals a most important cause of a new form is the isolation of a population at the fringe of its range. These fringe populations become separated, change in response to some slightly different set of conditions and eventually become different enough from the parent stock so that they cannot interbreed (i.e. a truly independent biological species). A shift in conditions may allow the descendent species to spread out and even displace its progenitor. Because the inception of a new species happens at the edge of a population, the chances of finding the actual site where it happens preserved in the rocks is rather low; the point of origin is nearly always somewhere else. The result in any particular rock section is a rather patchy mosaic of species that all seem to be related, and in some cases may record actual ancestors and descendants, but which can certainly not be read "like the pages of a book".

Added to this is another kind of change that can be recorded in sedimentary successions. This is a slow drift of change that occurs within a species through time, a shift in shape perhaps, or an increase in size, which cannot be explained by isolation of populations. In this case the ancestral species is transformed into the descendant and so does not really become extinct in the normal sense of the word. Some palaeontologists hotly dispute that this kind of change represents evolution at all, and there is a lot of hair-splitting about whether or not the descendant is really a different species. Regardless of the theoretical arguments, there are many rock sections in different parts of the world where this kind of change has been seen. It is of practical importance, too, because these kinds of

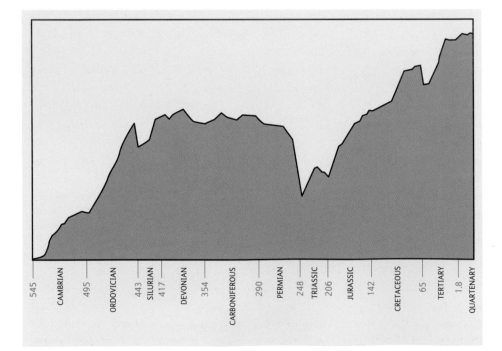

RIGHT: Numbers of major groups of marine organisms through geological time. After an early climb through the Cambrian, the numbers are rather uniform except at the periods of major extinction, especially at the Permian-Triassic boundary.

small changes often occur in the commonest fossils, and become the basis of stratigraphic zones for the precise dating of rocks by their fossil content (see right). This complicated introduction is necessary to understand the complexity of the fossil record. For example, if a new species arose by the kind of isolation of marginal populations that is important today, then we might expect the fossil record in any limited area to have gaps and jumps in it. In fact, the fossil record might not be so incomplete as has been thought; the jumps may be a natural phenomenon.

The generation of new species and the extinction of others has been a continuous process since the Cambrian. If one considers life in the seas, there was a steady growth in the total number of species during the Cambrian, and the number has remained high since then. But there have been a number of important times when extinction greatly exceeded the generation of new forms, and these were the times when whole groups, such as the rugose corals or the ammonites or the dinosaurs, passed from the world forever – mass extinctions. Very important mass extinctions occurred at the end of the Permian and the end of the Cretaceous periods, and it is no coincidence that they also define the end of the Palaeozoic and Mesozoic respectively. These were great biological crises that changed the faunas of the Earth. Here we have to invoke more powerful causes than the usual processes of change described earlier in this chapter, and these will be discussed later. But first it is necessary to describe the kind of small-scale evolutionary changes that are the bread-and-butter of the fossil record.

THE MOSAIC OF EVOLUTION

The evolutionary story is best preserved in the least conspicuous of fossils. More momentous changes have often taken place in sites where the fossils stand a low chance of being preserved; however, there are now fossils that serve to connect birds with dinosaurs, or fish with amphibians, even if some intermediate steps are still lacking. To see the record at its fullest one has to turn to sites where there is every reason to suppose that sedimentation has been continuous. One such place is in the deep sea, where the steady rain of plankton carries on regardless of the shifts in sea level that affect the continents, and the sea bottom is a graveyard for the rain of tiny organisms that die nearer the surface. Some of the small planktonic foraminifera record the kind of continuous changes that often evade the palaeontologist elsewhere. Minute species of the genus *Globigerinoides* change step by step into a perfectly spherical form known as *Orbulina*. The appearance of this spherical animal serves to define the base of the Mid-Miocene period. The same sequence of changes from the one species to the other has been found repeatedly in many sites all over the world. The cause of the change is a mystery, but whatever the reason the effect is of the greatest use in dating rocks of Tertiary age.

This kind of simple change can be matched by other examples from planktonic foraminifera, and it seems to be characteristic of planktonic organisms in general. Graptolites also show continuous changes of this kind. In the earlier half of the Ordovician, slender 'tuning fork' graptolites of the genus *Didymograptus* show a progressive change to stouter, longer forms, while the V-shaped species of *Isograptus* produce longer, more robust kinds of colonies. Ammonites also tend towards a continuous sequence of changes, sometimes with the multiplication of ribbing or with progressive changes in the outline of the shell. In these

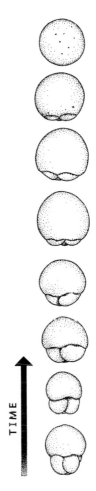

BELOW: Gradual transformation of the foraminifera *Globigerinoides* into *Orbulina*, from the Miocene, widely used for dating.

TIME

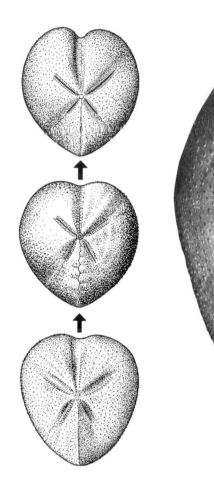

ABOVE: Changes which occurred in the heart urchin, *Micraster*, during the Cretaceous.

ABOVE RIGHT: This common heart-shaped Cretaceous sea-urchin, *Micraster coranguinum*. This specimen is preserved in its original calcite and measures 6 cm (2 ½ in).

planktonic or pelagic organisms it seems that whatever advantages were conferred on the species by the change were transmitted to the population as a whole.

Among bottom-living fossils it is perhaps more usual to find rather sudden jumps between one species and the next, probably because the species evolved by geographic isolation in the manner described previously. The sea urchin *Micraster* is one of the commonest fossils in the European chalk (Cretaceous), where its heart-shaped tests have acquired the common name of 'shepherd's crowns'. The chalk seems to have accumulated as a pure, lime ooze, largely formed by the microscopic remains of minute algae, and foraminifera. Although not a sediment of the open ocean, its record of deposition is remarkably continuous. Changes in the sea urchins record an evolutionary story connected with burrowing habits. As the sea urchins acquired the habit of burrowing deeper in the soft, chalky sediment their tests underwent a series of small changes: the entire animal became higher and inflated, the plates that make up the petal-like ambulacral area became more inflated, and the lip around the mouth became exaggerated. The series of species leading from *Micraster corbovis* through *M. cortestudinarium* to *M. coranguinum* record these changes, and the same species may be collected in the same order across much of Europe. They have become important in determining the zones of the chalk. In this case it is possible to correlate the changes in the fossils with their mode of life, by comparison with living sea urchins that show similar changes in relation to life habits. As the animals learned to burrow more deeply it became necessary to retain contact with the surface for breathing through a long tube, kept clear of sediment by extended tube feet. The points of attachment of these specialised tube feet became modified in a special way as the tube became longer. The *Micraster* species were not the only organisms to show slow, ordered, evolutionary change as the chalk was slowly deposited; large molluscs of the genus *Inoceramus*, as well as tiny foraminifera that make up much of the sediment, also changed, and all of their changes act as a chronometer for the passage of Cretaceous time.

The chalk is an exceptional deposit in its continuity and uniformity, without the kind of drastic change in sedimentary facies that introduce problems into more usual sedimentary sequences. However, even the chalk is not without its changes; there are beds where deposition temporarily slowed down or ceased, called 'hard grounds' because the sediment surface became crusty. Here a whole series of animals that are normally rare in the chalk become numerous, and unlike the sea urchins their evolutionary history cannot be deduced from careful collecting in the beds above and below, simply because they cannot easily be found there. With many fossil groups this is the normal state of affairs; fossil remains are found in certain special horizons, some of which become justly famous for the beauty of their preservation. They tell us much about the animal species concerned, but nothing about the history of the animal in the time immediately before or after. Usually the gaps between the fossil-rich horizons may be measured in millions, if not tens of millions of years, and they are often widely separated geographically. It would be a mistake to think we could stack these fossils up in their time relationships to give us a picture of how they changed, in the manner of the sea urchins. There are too many ambiguities in the gaps. The best that can be done in these circumstances is to look very carefully at the features they show, decide which ones were derived during the course of evolution, and construct a diagram to show the progressive stages in their pedigree (technically, a cladogram). For most fossils the construction of cladograms is the best we can do. As usual, it is a matter of debate how imperfect the record of a particular group of organisms is, and different scientists have differing views about how important the stratigraphic order should be.

In many rock sections it is observed that species remain relatively unchanged through a considerable thickness of strata, perhaps representing a million years of geological time. Then a new species closely related to the one in the underlying strata appears, quite suddenly, often replacing its predecessor, only to persist more or less unchanged through metres of strata in its turn. If evolution happened, so the argument goes, then it must be relatively fast compared with subsequent persistence of a new form. This phenomenon came to be known as 'punctuated equilibrium'. Although considered controversial when first proposed, such a scenario is probably no more than would be expected from a little consideration of how new species originate, and migrate and compete in nature. When a species moves into a new area and successfully out-competes a closely related resident, species the 'takeover' is surprisingly quick. In Britain we have seen the North American grey squirrel – aggressive, intelligent, and carrying a strange disease – drive out the native red squirrel during the course of a couple of human generations, which is a geological instant. Assuming the process continues, the red squirrel will disappear within a few decades, and the grey squirrel will continue unchallenged for a long time. Imagine if we had to read this story from fossils: what

OPPOSITE: Life in the Middle
Cambrian Seas. The large predator
Anomalocaris in the centre of
the picture belongs to an entirely
extinct group, while *Hallucigenia*
(see p. 156) has some distant
living relatives, as does the
priapulid 'worm' on the right.

we would see would be a long period with red squirrels, a very short (hardly visible) stage with both together, followed by (presumably) another long period of greys alone. It would appear as a punctuated range. If new species arise in small, peripheral populations, as has often been claimed, then any successful new form would invade and replace in much the same fashion as the grey squirrel. In fact, even as this is written, I understand a new mutant black squirrel is driving out some of the greys; *toujours la même chose*.

TIMES OF MAJOR EXTINCTION

Whether squirrel or sea urchin, the kind of evolution and extinction we have described so far has been on a relatively small scale. Even the traumas of an ice age, which may produce races or species able to cope with cold conditions (of which woolly mammoths and cave bears are examples), do not necessarily stimulate the mass extinction of whole groups of animals. Yet the extinctions that occurred at the end of the Permian and again at the end of the Cretaceous periods were of this kind. These were times of crisis for the whole fauna (and flora) through which only a few groups passed unscathed; many major groups of animals disappeared from the Earth forever. In some cases the extinction was preceded by a general decline in number of families, so that the trilobites had already been reduced to a few forms by the Permian, and their removal before the Mesozoic was only the coup de grâce. In other cases the termination was apparently more abrupt – the classic example being the extinction of the dinosaurs at the end of the Cretaceous.

These events seem enormously destructive and yet they allow the subsequent proliferation of other groups: extinction and evolution have been partners in shaping the modern biological world. While the dinosaurs occupied the dominant places in the terrestrial ecosystem there was no obvious opportunity for mammals to become the prevalent large herbivores, but once the dinosaurs were removed the rich herbage of the Tertiary was there for mammals to exploit. In both the Triassic and the Early Tertiary there was a short time lag after the preceding extinction events to allow recolonisation of the available ecological niches, but it was astonishing how quickly the faunas recovered to something like their previous diversity. The two major extinction events are the most pronounced of a large number of phases when extinctions were apparently higher than normal. At the end of the Triassic, for example, almost all the ammonites that had flourished in that period became extinct, a mere two groups surviving to give rise to the different kinds that populated the Jurassic seas. The end of the Ordovician saw the extinction of many of the characteristic trilobite families that make Ordovician fossil assemblages easy to recognise, and the same event affected brachiopods, nautiloids and graptolites. Mass extinctions are also thought to have occurred in the Late Devonian and in association with 'Snowball Earth' in the Late Precambrian. These extinction events define the ends of many of the geological periods; indeed it was probably the different gross characters of the faunas on either side of the boundaries that enabled the early geologists to recognise the different major periods to start with. The events at the end of the Palaeozoic and Mesozoic were probably the most dramatic. It has been suggested that the extinction of animal groups at these times may have been because of lower rates of evolution rather than particularly high rates of extinction, such that the normal rate of replacement of extinct animals by new species just did not take place. However, both events were associated with abnormal conditions in the biosphere as a whole; they were genuine times of crisis.

Sabatinca perveta, a moth belonging to the living pollen-feeding family Micropterygidae, in Burmese amber of late Cretaceous age.

nuclear war. Vast quantities of dust and sulphur were thrown up into the atmosphere, smoke poured from massive forest fires, blotting out the sun, and this killed the vegetation, which needs sunlight to photosynthesise. There was a similar catastrophic effect on plankton in the sea, which forms the base of all marine food chains. Because the largest dinosaurs were vegetarians they could cope neither with the loss of their food, nor with the cold weather. The carnivorous dinosaurs that preyed upon the grazers soon died out when their prey was no longer available. Survival was a matter of being able to tolerate low temperatures, or being able to enter a state of dormancy (as a seed, or a burrowing larva, for example), or perhaps just having the ability to eat a variety of foodstuffs. The small, warm-blooded mammals might have survived on a diet of scavenging insects. In the sea, crabs could have picked from the varied larder they still enjoy today.

This theory has a simplicity that makes it very attractive, if one can apply such a word to one of the greatest disasters in the history of life. It seems almost parsimonious to complain that there are some difficulties with it. The extraterrestrial impact seems very likely – but did it really cause the extinctions? In the first place it is claimed that the animal groups that became extinct, ammonites and dinosaurs, began to decline before the very end of the Cretaceous. If this were so it could scarcely have been in anticipation of an extraterrestrial impact. Second, it is hard to explain why some of the animals and plants that survived were apparently affected so little, for example insects and flowering plants. Cretaceous flowering plants are closely related to living ones, and most of the insects can be placed in living families, or even genera. Surely Cretaceous events would have affected the flowering plants as much as any group of organisms. One moth from the Cretaceous has living relatives that eat pollen, and there is no reason to suppose that its Cretaceous relative did otherwise. How could this moth have survived the darkened years? Similarly, colonial corals in the sea are exceedingly sensitive to light and pollution, yet they were not extinguished either. Already it seems necessary to introduce the notion of an area of the world that was not affected by the catastrophe to the same degree: possible, of course, but more complicated. There are also claims that the marine extinctions may not be as profound as has been thought. Another school of scientists point to the fact that the Deccan Traps in northwest India (formed from huge eruptions of basalt lava) coincide with the Cretaceous–Tertiary boundary. Could this have been an alternative source of the iridium anomalies and provide a more domestic explanation for the extinctions? After all, we know that volcanic eruptions have had profound climatic effects even within historical times. The catastrophists have countered that the type of volcanic eruptions which produced the Deccan Traps (continental flood basalts) are not of the explosive kind usually associated with gross climatic effects. They also claim to have found the site of the meteorite impact: Chicxulub on the Yucatan Peninsula in Mexico, the ancient home of Mayan civilisation. There have even been claims that evaporation of sulphur compounds at this site may have contributed to acid rain at the crucial interval. Although the jury is still out on the details, it does look as if the meteorite catastrophists are carrying the day. But palaeontology is full of surprises and it is not inconceivable that a new fact will turn the tables once again.

This short account of the Cretaceous–Tertiary extinctions shows just how complicated it can be to work out the causes of one of the most important events in the history of the biological world. It is obviously worthwhile finding an answer, sifting through all the contradictory evidence to find the real causes, even if the 'right' answer always seems to elude us. Whatever the final explanations for all the mass extinctions turn out to be, and each one may be different, it is undeniable that their effects have been the most important thresholds through which the fauna and flora passed to make the modern world.

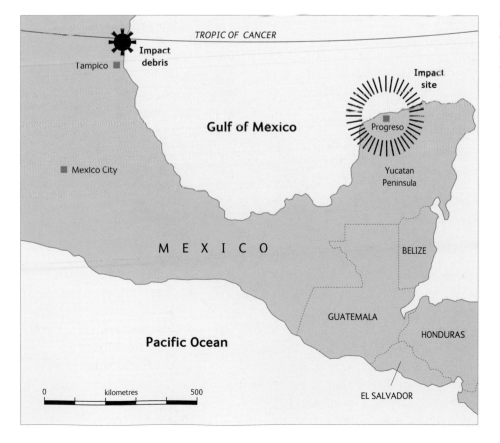

LEFT: Recent research suggests that an asteroid may have landed in the sea just off the tip of the Yucatan peninsula, Mexico, about 65 million years ago.

Human evolution

"The proper study of mankind is Man". The generous resources put into the hunt for the fossil remains of humans provide ready proof of Alexander Pope's aphorism. The search for palaeontological evidence bearing on human origins has probably occupied more newspaper space — and provided more scandal — than all other fossil discoveries put together.

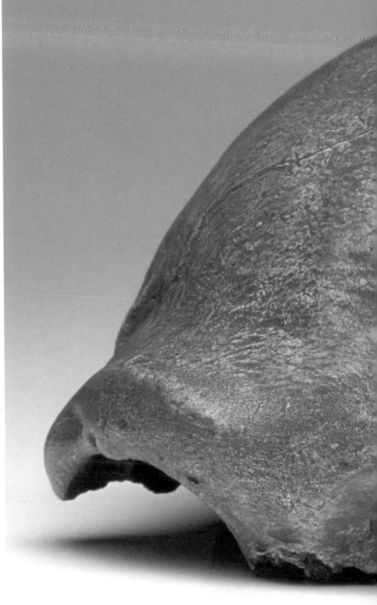

RIGHT: The first specimen of a Neanderthal, discovered with parts of a skeleton in the Neander Valley, Germany, in 1856.

Since Darwin published *The Descent of Man* in 1871, the idea that humans and the higher apes were closely allied has prompted a great deal of scientific investigation. The search was on for fossils that might flesh out the narrative of exactly how and where our own species, *Homo sapiens*, separated from its closest relatives, extinct or extant.

Like many scientific detective stories, it started simply and got more and more complicated. Critical fossils turned out to be very difficult to discover; they were assuredly very much harder to find than the brachiopods and ammonites that were collected in such profusion in the 19th century. The initial assumption was made that the transformation from ape to human might have been a simple series of step-by-step changes. Perhaps a similar notion still endures in the popular portrayal in cartoons of tiger skin-clad primitives looking discombobulated by the sight of a wheel. This simple model of human evolution led to the expectation that there would be a single transitional species – a 'missing link' to be found hidden somewhere in the rocks – which would carry an amalgamation of human and ape characteristics. It would prove our ancestry at a stroke. So eager were the scientific community for such a discovery that when the remains of an appropriate creature began to be unearthed in the south of England between 1908 and 1913 they were quickly accepted as the proof the world had been awaiting. Piltdown Man was a familiar name to 20th century scientists, presenting a mixture of higher ape and human characteristics. Somehow, it seemed appropriate that the remains should be discovered in Darwin's homeland.

Sadly, the British 'discovery' proved to be an elaborate forgery composed of bony bits of human and ape, carefully matched to look genuine. It was exposed as a fake when the different provenances of the several pieces were proved, and close inspection revealed that many bones had been dyed to simulate antiquity. One of those who had been seriously misled was the then head of the Geology Department of the Natural History Museum, Sir Arthur Smith Woodward. So it was rather appropriate that the man who was largely responsible for exposing the fraud in the middle of the last century, Kenneth Oakley, should be from the same institution. Most people now agree that the likely forger was the country solicitor who had made most of the 'discoveries', Charles Dawson – he had even had the 'fossil' named in his honour *Eoanthropus dawsoni*.

Fortunately for elucidating our history, even as Piltdown held its bogus sway, there were perfectly genuine finds made in other parts of the world. One former convention was to add 'Man' on to the place of discovery to identify a particular kind of fossil. One of the first such discoveries was made in 1856 when a skeleton was excavated from the Feldhofer Cave in the Neander Valley in Germany – so 'Neanderthal Man' this early human became. But with a brain case as capacious as that of modern humans, Neanderthal Man did not readily fill that role of a 'missing link'. Then, too, discoveries had been made in Asia, and these became Peking Man and Java Man respectively, whether or not they were originally derived from a female. More generally, and more scientifically, such early remains are all collectively referred to as belonging to 'hominids', the family group which embraces ourselves and our closest relatives (the term 'hominins' is often preferred in modern literature because 'hominid' may include chimpanzees and the gorilla too).

It soon became clear that early human history embraced a large part of the world, and, despite the rarity of our fossils, the story of human evolution was likely to prove rather complicated. The research of the last 50 years has amply confirmed this impression. Furthermore, in the early days of palaeoanthropology the dating of human fossils was highly contentious and fraught with inaccuracies. It was often impossible to say with certainty "this bone is older than that one". Some of the revolutions in our understanding were purely the result of technical advances. The development of radiocarbon dating allowed human remains – at least those younger than 50,000 years old – to be dated with unprecedented accuracy. This did not help with still older discoveries, of course, but other techniques were developed at the end of the 20th century that helped to put a date on even earlier bones – notably, the uranium series method, based on the different rates of decay of radioactive isotopes of uranium. This method in turn required the precision of new mass spectrometers that could measure almost inconceivably small quantities of isotopes. As so often in science, technical innovation went hand in hand with new discoveries in the field.

THE AFRICAN DISCOVERY

And there were plenty of discoveries. Although it is difficult to identify one particular find that was crucial in the story of unravelling human history, a case can be made that the single most important fossil in changing the focus of scientists was the recognition in southern Africa in 1924 of the skull of the 'Taung child' by Raymond Dart. He gave this animal the name *Australopithecus africanus*, and thereby also gave us humans a close relative that was not a member of our own genus, *Homo*. The skull was decidedly primitive in some aspects, for example in its brain capacity, and Dart himself still retained somewhat naive ideas about the fossil fitting neatly between great apes and humans. As so often in the story of human palaeontology, the fossil immediately attracted controversy, and some authorities of the day thought that Dart's child was more closely related to the gorilla or chimp (and some

BELOW: Discovered in 1924, the Taung child from South Africa was the first specimen to be called *Australopithecus*.

may even have preferred Piltdown as the 'ancestor'). Dart was eventually fully vindicated when more and more attention was directed at Africa as a source of fossils of humans and 'prehumans', but his rehabilitation took several decades. Although finds outside the African continent continued to be made, it became clear to a number of scientists that the cradle of hominin evolution was in that continent, and the search was on to flesh out the story with new fossil discoveries. Southern Africa continued to provide fossils (e.g. the locality at Sterkfontein proved very prolific), but the hunt spread more widely up into Kenya, Ethiopia and the Great Rift Valley.

There are good reasons why Africa was appropriate both for the evolution of humans and for their preservation as fossils. During the last 10 million years the climate of that continent has changed several times. At times of relatively high rainfall lakes proliferated and their shores abounded with game, and doubtless made good pickings for our ancestors —furthermore, lake sediments are good for preserving fossils. At other times the climate became more arid. Although there are variations on the exact scenario, many anthropologists seemed to agree that a change to drier conditions acted as a stimulus for evolution. It was argued that life on an open savannah with scattered trees and spaces to traverse would inevitably have brought our ancestors "down from the trees". Some of the important anatomical changes associated with becoming human followed plausibly: sight would have been at a premium in these more exposed circumstances; rising up on two legs would have allowed a clearer prospect to look for potential enemies or food; i#]ncreased capacity to communicate between the small bands in which our ancestors almost certainly lived would have favoured the more intelligent. A complication with this scenario is that we now know that the

ABOVE: Professor Dart and the Taung specimen of *Australopithecus africanus*. Dart was the first to recognise that it represented an early stage of human evolution. The specimen comprises part of a juvenile skull and mandible, and an endocast (petrified sediments that fill the cranium after death) of the right half of the brain.

main drying occurred after 2.5 million years ago, too late to have driven the evolution of bipedalism, so walking upright must have begun in or near the trees. But new evidence is emerging all the time. Periods of active volcanism in the Rift Valley also resulted in ashes that helped to inter the bones of these pioneers in the business of being human. By a happy concurrence, the same rocks can be dated rather accurately — so they provide a useful chronology of the seminal events in our evolution.

The squat pelvis of *Australopithecus afarensis* suggested that here was an animal capable of walking on two legs. This bipedal gait is fundamentally important to being human, and, along with features of skull and dentition, is an important reason to classify *Australopithecus* with *Homo* as our closest related genus. The conclusion from anatomy was triumphantly confirmed when fossil footprints of *Australopithecus* were found at Laetoli in Tanzania, being

This trail of hominid footprints fossilized in volcanic ash was found during an expedition at Laetoli, Tanzania, in 1978. The trail probably belongs to *Australopithecus afarensis* and dates from 3.7 to 3 million years ago, and shows that hominids had acquired the upright, bipedal, free-standing gait of modern man by this date.

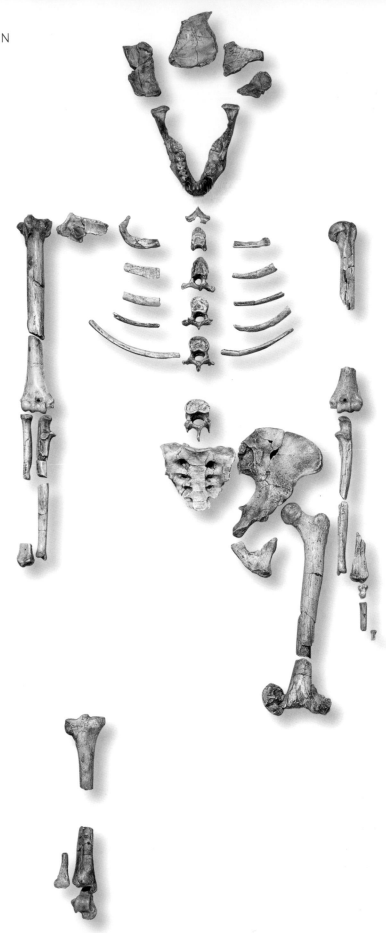

RIGHT: Partial skeleton of
Australopithecus afarensis, a
primitive African hominid that
lived 3 to 5 million years ago,
nick-named 'Lucy'.

the very tread of our remote predecessor's feet on soft volcanic mud preserved before it could be eroded away. Since our closest *living* relatives are the chimpanzee and bonobo (*Pan*), which are still resident in Africa, the branch that separated our human line from that of the great apes must go back before the first *Australopithecus* fossil, which is certainly more than 4 million years ago.

As more fossils have been found, the story of human origins has started to look more and more complicated. There are primitive forms from the period 5–7 million years ago, each assigned to a different genus, *Orrorin* (from Kenya), *Sahelanthropus* (Chad) and *Ardipethecus* (Ethiopia), but it is unclear how human-like any of these really were. Additionally, many different species of *Australopithecus* have now been named – there are at least seven species generally recognised as this is written. There are still many differences of opinion about where one species ends and another begins. Obviously all seven cannot be on the *direct* line to *Homo* and us, which implies that there were lines of evolution that simply petered out. However, the species of *Australopithecus* do divide quite naturally into two groups. The robust australopithecines are, as the name implies, hominins with very powerful jaws, small brains and flat faces – a popular name for one species was 'Nutcracker Man', for reasons that should be obvious. These robust species prospered in drier conditions, where they could tackle tough seeds and similar foods. Many anthropologists think that these 'ape-men' are different enough to be placed in their own genus, while very few anthropologists believe that these rather inelegant creatures were on the main line of human evolution. They diversified for a while and then died out. Most anthropologists would rather point to another group of species with a more delicate frame as the likely ancestors of humans – these 'gracile' *Australopithecus* species include *A. africanus* and *A. afarensis*. The famous and rather complete skeleton known as Lucy discovered by Donald Johansen and his colleagues in Ethiopia in 1974 belongs to *A. afarensis*. It is commonly assumed that the ancestry of our own genus is rooted in a hominin resembling *A. afarensis* or *A. africanus*. Thus Raymond Dart's original discovery is still a contender for an ancestor of *Homo*.

EARLY HUMAN

As with *Australopithecus*, so also with *Homo*. Not many years ago, there were very few named species of *Homo* – now there is a whole clutch of them. The original idea of 'links in a chain' leading to modern humankind has been replaced by a view recognising a number of branches that did not give rise to any survivors, and a rather special lineage that led to the whole panoply of the human race in all its diversity of colour, language and creed: that is to say, *Homo sapiens*.

The earliest history of *Homo* is entirely African. One of the oldest species recognised from fossils that can be attributed to *Homo* is *H. habilis* – the 'Handy Man' of early accounts. This species' remains go back as far as 2.3 million years according to some authorities. The brain capacity of this early member of our genus was only 30% more than that of *Australopithecus africanus*. But primitive though this early human was in many features, it was also able to manufacture primitive tools – and the common use of tools is one indication that the species is closer to us than to its presumed *Australopithecus* ancestors. *H. habilis* was found by Louis Leakey and his team in the Olduvai Gorge in northern Tanzania, another locality where an ancient lakeside combined with volcanic ashes to yield an exceptional

RIGHT: A collection of pebble
tools discovered at Olduvai Gorge,
Tanzania. Fossils found with
the tools suggest it was *Homo
habilis* that made the tools.
They were not sophisticated in
today's terms, but to make them
would have required 'human'
adaptations of the hand.

record of early humankind. The Leakeys are probably the only dynasty in anthropology. Louis did much to establish what he called the "cradle of mankind" in Olduvai in the 1950s and subsequent decades; his wife Mary was almost as redoubtable (she actually found the skull of 'Nutcracker Man' there). Richard, their son, has not only followed in the tradition but outstripped his parents in sheer number of discoveries of hominin remains. During the 1970s Richard Leakey, with his colleagues, found the remains near Lake Turkana, Kenya, of what some anthropologists recognise as a separate early species of *Homo*, *H. rudolfensis*, which elsewhere dates back as far as 2.3 million years old, and another species, 1.8-1.4 million years old, *H. ergaster*, which has been claimed as widespread in Africa, and which is associated with prolific tool making and, quite possibly, fire. It would be fair to say that not all anthropologists agree on the species number and status of these early hominins, but new discoveries since the Leakeys' groundbreaking finds continue to add details to an increasingly complicated early picture of our own genus.

Many anthropologists regard *Homo ergaster* as no more than an early form of *H. erectus*, whose fossils have now been widely collected in both Africa and beyond. In fact, the type specimen of *H. erectus* is one of the specimens of 'Java Man' collected before the great explosion in African discoveries. *Homo erectus* ('Upright Man') might seem a little unfortunate as a species name, now that we know that ancestral australopithecines were also capable of upright gait. However, in many of its characteristics *H. erectus* scales rather well as an intermediate between *H. habilis* and *H. sapiens* – having a brain capacity about 70% of that of our species, for example, while the stature difference between the sexes is somewhat

LEFT: A *Homo erectus* ('Java man') cranium from the Sangiran site in Java, showing the exaggerated browridge and thick skull bones.

LEFT: One of the skulls of *Homo erectus* from Dmanisi, in the Republic of Georgia, western Asia, before it was removed in 2001. It did not show all of the expected features of other *H. erectus* specimens suggesting that some populations of *H. erectus* could be different from others.

greater than ours. *Homo erectus* was a skilled toolmaker, in some cases flaking both sides of a variety of hand axes, and adept at finding sources of flint or obsidian as a raw material. Fire had been used to cook meat. We do not know to what extent *H. erectus* was able to speak, but it is very likely that the species lived in small cooperative groups in which good communication would have been at a premium.

Now that it is established that the root of human evolution was in Africa, it follows that *Homo erectus* must have migrated from its original homeland, eventually penetrating as far eastwards as the Indonesian archipelago. This implies a great measure of adaptability and perhaps even a sense of adventure. The route out of Africa is usually considered to have been through the Middle East and would have been favoured at times when the climate was wetter; the Great Rift Valley may have provided a natural corridor for the movements of human populations along with the game they preyed upon. It is clear from associated bones that *H. erectus* was an effective hunter, but that a diet of meat was probably supplemented by roots, vegetables and fruits, much like surviving hunter-gatherer communities. This encouraged a style of dentition much more akin to our own. 'Java Man' lived about 700,000 years ago. Some of the most complete specimens of *H. erectus* from Africa are considerably older. For example a nearly complete skeleton found by Richard Leakey's team in 1984 from Nariokotome, Kenya, is dated as 1.6 million years old (other similar fossils are as old as 1.75 million years in Africa and the Georgian site of Dmanisi, western Asia).

For a while, after the middle part of the 20th century, a rather curious theory claimed that *Homo erectus* evolved into modern humans over its whole range as a kind of gradual transformation. This idea was unlike the way most new species appear within relatively small populations that eventually displace (or even live alongside) their ancestor. The 'multi-regional origin' has largely fallen into disfavour as another, and ultimately African origin, for our species has gained currency.

Homo heidelbergensis is in many ways a close anatomical forerunner of *Homo sapiens*, and indeed for many years was classified with our species under labels such as "archaic *H. sapiens*". African fossils as old as 600,000 years are now assigned to this species, which was almost certainly derived from *H. erectus*, although its cranial capacity even overlapped that of modern humans at the upper end of its range. From the point of view of its anatomy it provides a bridge between *H. erectus* and *H. sapiens*. 'Heidelberg Man' was widespread in Europe, even crossing into Great Britain before the English Channel had been cut to leave a rich legacy of stone tools (and prolific evidence of a hunting lifestyle) at Boxgrove in the south of the country.

But what of 'Neanderthal Man', that human relative discovered nearly a century before most of the crucial fossils were recovered from Africa? I should mention that some authorities recognise yet another species, *Homo antecessor* (c. 0.8–1.2 million years old), as a common ancestor of both Neanderthals and the modern human line, which is known from fossils from cave deposits in Spain, for example. It would be fair to say that not all anthropologists accept the reality of this species at the moment. As for Neanderthals, they were short and squat, robustly built, with large and projecting nasal openings on the skull, which almost certainly betokens big noses. Many of the skeletons recovered show evidence of healed broken bones, and it is likely that Neanderthal life was "nasty, brutish and short". The brain of the Neanderthal was as capacious as our own, although the brow tended to slope back more. It is likely that some of these tough stocky people were adapted to life in the frigid conditions of the last ice age, when they hunted mammoths, reindeer and other mammals that prospered in those challenging conditions, while others lived in more equable conditions around the

ABOVE: *Homo neanderthalensis* cranium Three-quarter view of an adult female Neanderthal cranium discovered at Forbes Quarry, Gibraltar. Its discovery was announced by Lieutenant Flint in 1848 and it is believed to be 50,000 years old.

LEFT: Neanderthal artefacts from Gorham's Cave, Gibraltar. These stone tools were critical in carving meat from bison, reindeer, and other large mammals.

northern Mediterranean. As the ice-sheets waxed and waned, so the Neanderthal's range expanded or contracted in harmony with the climatic shifts. Over tens of thousands of years they occupied much of Europe and Asia. They were, in their own way, specialists, and for a time very successful ones.

The naming of Neanderthals provides an interesting example of what has happened to the portrayal of human evolution as taxonomy has changed. Neanderthals have been regarded, at various times over the last 40 years, as a race of *Homo sapiens* adapted to Arctic conditions, or a subspecies of the human species, or, most recently, as a separate species, *H. neanderthalensis*. They appeared more than 400,000 years ago.

But these relatively recent fossil relatives of ours have yielded chunks of fossil DNA from exceptionally preserved bones – and DNA sequencing has revealed how like humans they were at the molecular level (see Chapter 9). In 2007 it was announced that Neanderthals even carried a gene for red hair. There has been much debate over one particular fossil that may (or may not) show evidence of cross-breeding between Neanderthals and early *H. sapiens*, which were contemporaries and seemingly, at times, neighbours. If that really happened, then they were assuredly not far apart genetically. Neanderthals died out about 28,000 years ago. Some authorities believe that their demise was hastened by our own species; other authorities prefer to believe that Neanderthals succumbed to environmental shifts to which they were ill adapted. Although they were well equipped to cope with Eurasian conditions during glacial times, they may have been finally out-competed by our species during the very unstable climates that prevailed between 50,000 and 25,000 years ago. But it would be rather rash to assume that we have heard the last about the classification and extinction of 'Neanderthal Man'.

ABOVE: Tony Djubiantono from the National Centre of Archaeology, Jakarta, Indonesia, holding a *Homo floresiensis* skull, found in 2003 in the Liang Bua cave, island of Flores, Indonesia. *H. floresiensis* is thought to have become extinct 12,000 years ago, and so co-existed with modern humans, *Homo sapiens*.

The interesting thing about palaeontology is that it is always capable of springing a surprise. The African story seemed to be becoming quite well fleshed out. But in October 2004 a totally unexpected discovery was made half a world away from Africa. In the small tropical island of Flores on the Indonesian archipelago the passably complete remains of a diminutive human being were discovered. This little person probably less than a metre tall was rather rapidly christened 'the Hobbit' by the media, but it received the less emotive scientific name of *Homo floresiensis* in the scientific press. The small hominin seemed to have rather a low cranial capacity – even smaller than that of *H. erectus*. Astonishingly, it appeared to date from rocks that were only about 15,000 years old – so if it were a descendant of *H. erectus* it must have long outlived its ancestor and been a contemporary of our own. Since there are stone tools on Flores going back some 800,000 years there is every possibility that this human originated by long-term isolation, another example of evolution in a small and separated population. But once again the latest discovery immediately became the subject of controversy. Some observers claimed that *H. floresiensis* was no more than an aberrant human – even a 'cretin'. However, as more specimens are examined it really does look as if *H. floresiensis* is a very distinctive species. Considering its small size, its feet are large and flat, lacking a long 'arch', and some of its wrist bones seem as primitive as those of

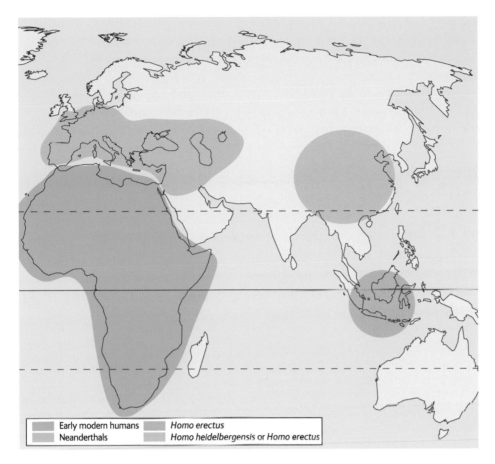

When modern humans evolved in Africa 150–200,000 years ago, Neanderthals were still living in Europe and western Asia, and *Homo erectus* or other hominin species were living in eastern Africa. After this point, modern humans began to spread out from Africa to replace other human species.

Early modern humans	*Homo erectus*
Neanderthals	*Homo heidelbergensis* or *Homo erectus*

The first appearance of *Homo sapiens* in the Middle East may have been temporary, as Neanderthals occupied the region after the first modern humans. At 50–60,000 years ago, *H. sapiens* spread out of Africa again, replacing other hominin populations.

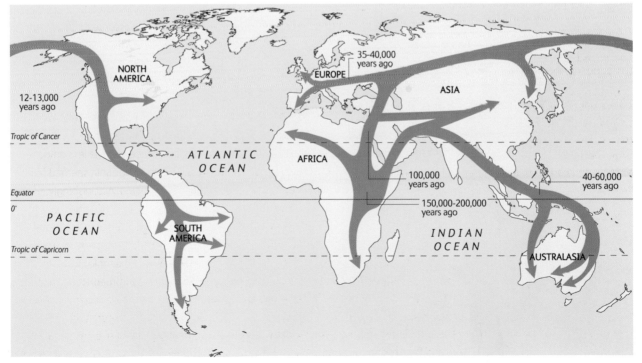

chimps or *Australopithecus*. These kinds of differences would not apply if the little men were merely odd human beings. As we have seen it is as well to be cautious in matters of human evolution, but it does seem very likely that as hominins evolved towards *H. sapiens* elsewhere in the world, an isolated population of *H. erectus* embarked on its own evolutionary adventure in one small corner of the tropics. It is even possible that *H. floresiensis* is more closely related to *H. habilis* and a still earlier phase of human evolution. The result was one of the least anticipated finds of the last century. Nonetheless, the 'Hobbit' remains a side branch in our history.

MODERN HUMANS

As for the main story – it is back to Africa once again. Modern humans – *Homo sapiens* in the current, restricted sense – originated in Africa perhaps 200,000 years before the present. Even a decade ago this statement might have been controversial, but today the great majority of scientists would agree with it. The evidence has come from several directions, all pointing to the same conclusion. As with other hominins, fossils of the very earliest modern humans are rare; later in our history they become much more numerous. Nonetheless, a series of finds in Kenya, and in several localities in Ethiopia, such as Omo and Herto, have shown that humans almost indistinguishable from us were widespread across the African continent 150,000 years ago. From this nucleus *H. sapiens* spread around the world, the greatest diaspora of any species in the history of life.

The appearance of modern humans is recorded in changes in stone tools, which are much more commonly preserved than body fossils. Although use of tools is deeply embedded in the history of apes and humans (Jane Goodall long ago demonstrated how chimpanzees

BELOW: Pierced seashells could be threaded to make necklaces, like these *Nassarius kraussianus* shell beads found in Skhul Vanhaeren d'Errico, Israel.

intelligently employ simple tools), only the line leading to humans developed what might be termed technology. *Homo erectus* and *H. heidelbergensis* produced simple but effective stone hand axes, but their 'industry' hardly changed for a million years. By contrast, *H. sapiens* increased both the variety and sophistication of tools, introducing fine flaking along stone tool edges, for example, that refined their cutting abilities, and specialist tools in bone, antler and ivory for particular purposes, such as harpooning fish. The succession of stone tool 'industries' was widely used to subdivide human history into Palaeolithic, Mesolithic and Neolithic times, based upon increasing sophistication of the tools produced, and this system still has its uses in regional archaeology. Humans were apparently also artists from early times. There is some argument about the first objects made with artistic intent, but pierced seashells that could be threaded to make a rudimentary necklace are known from Israel, North and South Africa and dating between 75,000 and about 100,000 years ago, across the whole known range of early *H. sapiens*. Much better known are the splendid rock paintings of animals which adorn the dark interiors of many limestone caves in Spain and France: these are much younger, 40,000 years old at most, and made long after humankind had left the African cradle. Nonetheless, it is not inappropriate to define our species as the hominin that makes art.

Further lines of evidence supporting an African genesis were derived from studies of the human genome. Even in the 1960s Luigi Cavalli Sforza had shown how basically similar the various races of humans were, and had mapped an evolutionary tree that suggested the greatest genetic diversity in Africa, which was likely to be the 'stock' from which other branches of the human tree arose. More sophisticated methods followed, which produced maps of the evolutionary history of humans, showing (to cite one example) that North American native peoples and eastern Asian peoples formed a single group – and therefore that the North American continent had likely been colonised through the Bering Strait from what is now Siberia. African peoples consistently occupied the basal branches on the tree, showing greater genetic variety than the rest of the world added together, and a famous study of the mitochondrial DNA inherited through the female line implied that we humans had descended from a particular group in Africa, one that the 'molecular clock' – which is described in more detail in the next chapter – dated at about 140,000 years ago. It was all exciting confirmation of our comparatively recent origin in Africa, which must have been followed by rapid exit (or exits) at favourable times to colonise the rest of the world.

Details of that spread are complex, and still controversial. There is no question that early members of our species were able to talk among themselves; as the tribes dispersed, so different languages arose. Although all languages share the same deep structure they have separated into a thousand different tongues over tens of thousands of years. It is evident that language evolves far faster than does our biological self, although the 'racial' characteristics that separate Caucasian from Oriental, and both from Australian Aborigines, were doubtless forged during the human diaspora. Some of the regional differences that are present today are obviously adaptive: dark, pigmented skin is helpful in tropical climates, and a pale complexion works better at high latitudes; peoples living in arctic climes tend to have shorter, rounder stature to conserve heat. Nonetheless, the genetic difference between all the peoples of the world is relatively slight.

It is plausible to believe that human tribes may have spread rather rapidly along seashores – and middens of discarded shellfish survive today to tell of this. The intelligent biped would have hunted and gathered opportunistically along the way. Since this was also the time of the great Ice Age, when glacial conditions reached maxima at northerly latitudes, sea level fell globally and exposed wide plains along the edges of continents, in the process also making land bridges. One of these permitted a passage through from Asia into what is now Alaska (which at times remained ice-free) about 13,500 years ago: the Americas were colonised from there. The first Australians may already have reached that continent 45,000 years ago, if the radiocarbon dates from old fires are to be trusted (presumably bypassing the 'Hobbit' in the process). Humankind was, from the first, an adventurer.

There is no doubt that *Homo sapiens* brought change wherever they went. This change may not have been to the benefit of animals previously unaware of the adventurous (and hungry) biped. The extinction of many large North American mammals – animals as varied as horses and giant sloths – coincides quite closely with the appearance of humans in that region, along with their finely crafted stone tools, which are often discovered associated with butchered bones: the Clovis culture. We know for sure that humans were capable of wiping out edible prey because of the disappearance of that giant, flightless – and presumably tasty – bird, the moa, shortly after Polynesians arrived in New Zealand – the last large landmass to be colonised (excluding Antarctica, of course). Did humans play a similar role much earlier in the destruction of the large mammals of North America (and then South America), even including predators like sabre-tooth cats? And were they responsible in Australia for exterminating large marsupial herbivores, the skeletons of which have been found dating from a time after humans had appeared 'down under'? Many scientists favour the view that humans were, indeed, the prime cause of these important extinctions. But it is also undeniable that there were major climatic oscillations throughout the same period, as the great ice-sheets waxed and waned in the northern hemisphere. Even today, as one flies over the outback of Australia, huge stabilised dune fields are easily visible, which date from a time when the continent was thoroughly dry, even by today's standards. The different claims of climate change *versus* human gastronomic intemperance in causing extinctions have taken up many pages in scientific journals and books, and it is easily possible to imagine the effects of almost any combination of both together. In arid times, for example, populations of large herbivores would decline quite naturally, and humans would get even hungrier, and that could add up to a lethal combination. Whatever the true history, it is something we need to ponder carefully at the present day when we are probably pushing more animals to the brink of extinction than at any time in our prehistoric millennia.

ABOVE: Did humans or climate change or a combination of both, wipe out the sabre-toothed cat, *Smilodon fatalis*? The skull of an extinct sabre-toothed cat that lived about 15,000 years ago in North America.

The history revealed by fossils merges insensibly into the province of archaeology. Human populations grew relentlessly as the ancient hunter-gatherer lifestyle was transformed. Domestication of animals and plants allowed for a settled life. Wheat probably came into cultivation in the Middle East 10,000 years ago and rice in the Far East; maize was 'tamed' independently in South and Central America. Agriculture allowed for a predictable surplus of food for perhaps the first time in human history. Settlement encouraged villages, and then towns, which left their own records measured in artefacts produced by a burgeoning technology. Some 6,000 years ago in the Middle East and India the settlements became large enough to support ceremonial centres, and the recorded parade of kings and queens began. What is dubbed civilisation marked the end of the palaeontologist's long period of concern and the beginning of the compass of the historian.

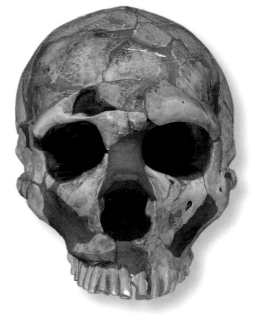
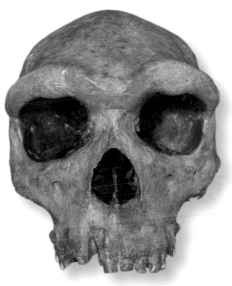
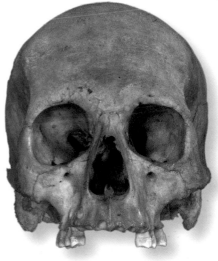

LEFT: Although it is possible to line up skulls like this through time, the actual pattern of human evolution is more complex. Top left clockwise: *Homo erectus*, *H. heidelbergensis*, *H. sapiens* and *H. neanderthalensis*.

Fossil DNA

The title of this chapter might seem surprising – and 30 years ago, would not have been considered a possibility: such is the march of science. The great strides that have been made in understanding the structure of the genome have impacted on every corner of biological sciences, and palaeontology is no exception, for all that the subjects of study are largely extinct.

RIGHT: A northern Siberia spring, 1.8 to 11,000 years ago with woolly mammoths, horses and a wolverine roaming amongst larch and pine, and sedge and cotton grass. DNA of frozen woolly mammoths can tell us about the relationships of the extinct mammoths to today's animals.

DNA unifies all life, so no matter how ancient the organism it is certain that it carried the genetic code, the famous long 'double helix'. The study of genes can provide another window into the past.

DNA itself can be preserved as a 'fossil' under exceptional circumstances. It is actually rather a delicate molecule, easily destroyed by excessive heat, and it readily decays. It is not surprising that DNA apparently does not survive in ancient rocks. 'Fossil' DNA is almost invariably split into chunks of varying length, and even then is present only in minute quantities. However, modern techniques can 'breed' such tiny pieces and replicate them to the point where there is sufficient for analysis. Such techniques will be familiar to many readers because they are often used in forensic science to identify criminals on the basis of a few hairs or incriminating stains. The problem with this technique is that contamination can happen very easily. A piece of bone may contain bits of DNA from all kinds of organisms other than the one under the microscope – fungi, or scavenging beetles, or the ubiquitous bacteria. These bits can all get 'magnified' too.

At one time there were claims that DNA could be recovered from very ancient fossil material such as amber. The idea that amber might act as a 'time capsule' was certainly an attractive one, and results were published reporting discovery of fragments of bee DNA from a fossil in amber. Unfortunately, when attempts were made to replicate this discovery under completely sterile conditions they were not successful. It began to seem likely that a small amount of 'rogue DNA' had found its way into the original sample – which was not unlikely in a laboratory where DNA sequences of other insects had been previously prepared. It is now considered unlikely that informative chunks of DNA can survive from deep geological time measured in millions of years.

However, under certain conditions identifiable lengths of the giant molecule can survive in samples many thousands of years old. Not surprisingly, the relevant work started on extinct birds whose history was known in some detail. The flightless dodo was exterminated on Mauritius in the 17th century. DNA was fairly readily obtained from bone samples kept in museum collections. Analysis of these samples was able to prove that the dodo was actually a giant pigeon – rather than, say, being related to ostriches. The DNA in question was sequenced, which means that the complex sequence of nucleotide bases along a given chunk(s) was determined. The longer the sequence the more distinctive it is likely to be. Sequences in particular genes are rather conservative – so if the sequence obtained from the dodo proved more like that of the same gene in the pigeon than that of the ostrich it follows that the dodo is most closely *related* to the pigeon. It is possible to extend the study further. For example, if the same gene were sequenced from a reptile, the chances are that sequences from all three birds would be more similar to one another than to that of the reptile, providing evidence that birds originated later than reptiles – and that they form a discrete biological group. So gene sequences can tell us about the history of descent in a way that often complements evidence from fossils. They provide another kind of 'fossil record'.

Maoris colonised New Zealand in about AD 1250 and by the time the first Europeans arrived a few hundred years later there were no surviving examples of another giant bird, the moa. Having evolved in isolation, moas grew to a great size (up to 3.5 metres or more than 11 feet high) and made ready prey for greedy humans when the latter finally made their way to these remote islands. DNA surviving within the bones of moas tells us that these species *were* related to the ostrich – and indeed to another extinct flightless species, the elephant bird (*Aepyornis*) of Madagascar, which also became extinct in historical times; the giant egg of this bird is the size of a rugby football. Humankind, it seems, has a lot to answer for in exterminating vulnerable species. Some recent molecular work suggests that the moa species were fewer than had been thought, but displayed quite striking differences in stature between cock and hen birds.

The extension of ancient DNA studies deeper into the fossil record proved productive. Bones of ice age animals that had taken refuge in caves – or been carried there by predators – yielded useful fragments of DNA, particularly if they had remained relatively dry. The prospect

LEFT: Skeleton of the extinct heavy-footed moa bird, *Dinornis elephantopus*, found in New Zealand.

of looking back to the early days of humans beckoned. By 2006 it was announced that as much as a million base pairs of Neanderthal Man's DNA had been sequenced (that sounds a lot, but is still only a tiny piece of the genome) from a specimen more than 30,000 years old. If these results are correctly interpreted, then *Homo neanderthalensis* may have been less close to *H. sapiens* than we thought – though it did share a version of our gene for red hair.

DNA tends to accumulate mutation changes through time. Many of these are not adaptive changes – the kind that might, for example, change plumage colour in a bird – but rather 'neutral' changes that don't do much except get handed on to subsequent generations. Hence accumulated changes in a given segment of DNA can be used as a kind of clock. If two species split apart from a common ancestor, the timing of that split can be estimated by the changes that have gradually become incorporated within the appropriate piece of DNA. The 'molecular clock' is not without scientific problems, but often provides the only evidence for some deep-seated event. In the case of our Neanderthal relative, the 'clock' suggested that this species separated from the main line of human evolution as much as 500,000 years ago. This was longer ago than might have been anticipated if Neanderthals had appeared close to the origin of *Homo sapiens* in Africa.

DNA turns up in some surprising places. Permafrost is found at high latitudes in areas like Siberia where the ground never thaws out completely. It has recently been discovered that if cores are drilled through the permafrost layers DNA can be extracted from sampling the frozen ground; the record goes back many thousands

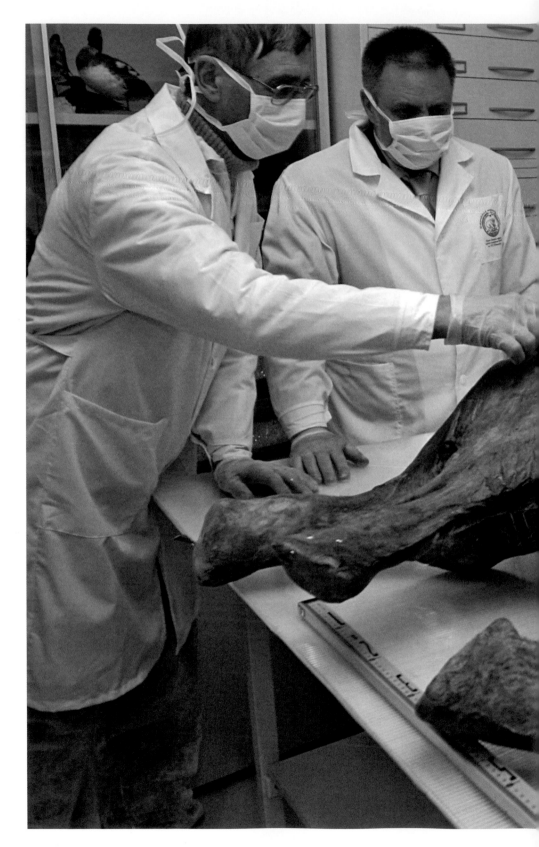

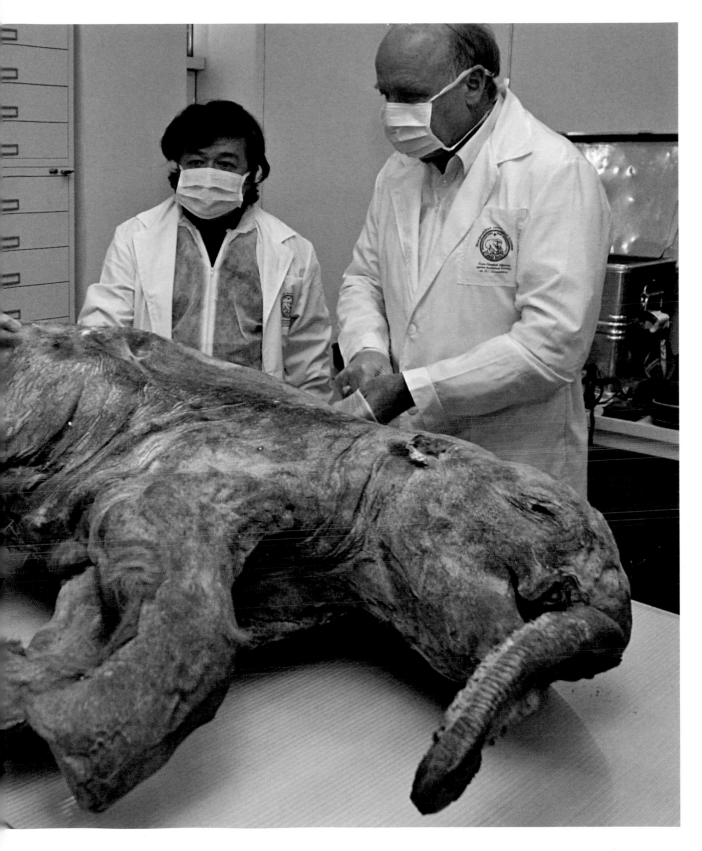

of years. Urine, faeces and beetle carapaces are all possible sources of the 'signal'. DNA from woolly mammoths was known previously from deep-frozen carcasses that still thaw out from the permafrost from time to time. Mammoth and musk ox DNA can also be recognised from the permafrost record, along with that of other mammals. The quantity and quality decreases as the layers go further backwards in time. At the moment it seems that about 60,000 years is the upper limit for the survival of usable amounts of DNA in the fossil record.

Molecular information has had other important effects on palaeontology. Evolutionary trees are now routinely constructed on the basis of relative similarities of gene sequences. Species having similar sequences are likely to be close together as 'twigs' on the tree; more widely shared similarities might indicate the 'trunk'. In some cases these trees are similar to those constructed long ago from fossils – but in other cases they are different, which has prompted creative thinking on the part of both palaeontologists and molecular biologists. Branches on such trees are dated by the first appearance of any fossil belonging to that particular branch. So palaeontology has a very specific relevance to molecular biologists: fossils help to date evolutionary rates. When an evolutionary line splits in two, both branches must, of course, be contemporaries. If the fossil dating of one branch is much older than its 'twin', this provides an indication of what is 'missing' from the fossil record of one branch. On the 'deficient' side there must have been fossils of which no trace has yet been recovered (this is evocatively termed the 'ghost range'). Ever since Darwin the imperfection (or otherwise) of the fossil record has been a matter of discussion – now there is a way to measure it. By totting up the 'ghost ranges' the scientist can get a measure of how fully he or she has sampled the species that once existed.

There are many organisms for which it is obvious that the fossil record will be woefully imperfect: tiny soft-bodied 'worms', mushrooms, bacteria or butterflies come to mind. Molecular evolutionary trees are practicable for such organisms, and they have frequently revolutionised their classification. However, what molecules can never tell us is what the intermediate stages in the process of evolution actually looked like – and fossils can. Recall that human evolution was driven by walking on two legs – big brains came later. We could not have inferred that without fossil evidence. Many interesting fossils show us how the land was colonised by plants in the Silurian, and this is at least as informative as learning about the inter-relationships between living primitive plants like mosses, liverworts and ferns: there is still no substitute for history. Nowadays molecular evolutionary biologists and more traditional experts work happily alongside one another in the common cause of unravelling the history of life.

'Molecular clocks' also have stories to tell. If there is sketchy evidence from actual fossils the dates provided by this method suggest what still remains to be discovered. To give one example, modern birds (Neornithes) have a rather sketchy fossil record – which is not surprising given the fragility of their bones. The 'molecular

LEFT: Model of one of the earliest tetrapods, or animals with backbones, capable of living on land. This *Acanthostega* lived during the Late Devonian of East Greenland. It had seven fingers, eight toes and a fish-like tail.

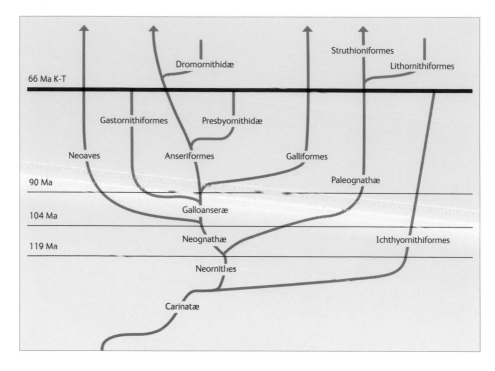

clock' based on the evolution of living birds suggests that the main evolution of these birds
had happened already before the extinction of the dinosaurs in the Cretaceous period. It
is now established beyond reasonable doubt that birds were descended from dinosaurs –
and that there were a variety of birds in the Cretaceous – but the fossils are dominated by
archaic kinds, and there are few Neornithes. Traditional palaeontologists hold to the idea that
modern birds did not really get going until after the major extinction event at the end of the
Cretaceous – after the dinosaurs and flying reptiles had disappeared from the scene. The
'clock' suggests on the other hand that there are new finds to be made in Cretaceous strata.
Time will tell which scenario is correct – but the evidence from molecules does suggest we
should keep on looking.

Genes make things happen during the development and growth of every organism; they
determine front from back, make limbs or eyes. During long stretches of geological time
new organs evolved – like legs from fins. So when we look at a series of fossils that bridge
this transition from water to land we are also looking at important changes in what is known
as the *expression* of genes. This particular sequence of fossils has been the subject of
intense scrutiny over the end of the 20th century and into the present one. Discoveries in
Devonian rocks in Greenland and Arctic Canada have made this one of the best studied of
the important transitions in the history of life. Unlike, say, the evolution of the insubstantial
eye, limb bones can record a full record because they are strong and fossilisable. So now a
series of fossils that record the change from fish to land-living tetrapod are known, and when
another is discovered it always creates a stir. Very early land-living animals from the Devonian
rocks had primitive limbs – some of them, like *Acanthostega*, even bearing more digits than
the usual five – alongside a number of retained fishy features. As recently as 2004 a fossil
was discovered in the Canadian Arctic from 375 million years old rocks that was 'a fish with a
wrist' – almost the ideal transitional form between water and land. It had what the discoverer

described as 'a shoulder, elbow and wrist composed of the same bones as an upper arm, forearm and wrist in a human' and was called *Tiktaalik* (see p. 122).

The exciting thing about fossils like *Tiktaalik* is not just the graphic confirmation they provide of the reality of evolution. They also show how skeletal development changed at a crucial time. Development of individuals can be studied under the microscope by the classical techniques of embryology: watching the fates of cells as they divide and prosper. Nowadays this is supplemented by new techniques that can show how this unfolding is controlled by genes that 'switch on' (or express) at certain stages in development. Finding these ancient fossils is like having a time capsule to see how an organ such as a limb transforms as evolution proceeds. Discovery of an ancient fossil serves to bind together the separate disciplines of embryology, palaeontology and the latest in molecular biology: we live in interesting times.

ABOVE: *Tiktaalik rosae*, an air-breathing amphibian with a skull, neck and ribs similar to those of early tetrapods, and fish-like jaw, fins and scales. The rocks in which the fossil was found were formed in shallow tropical rivers and lagoons

Uses of fossils

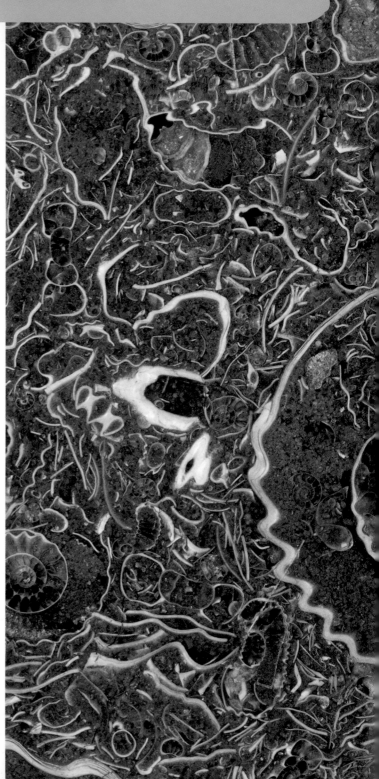

Every time you drive a car you are able to do so because of the photosynthetic activity of plants many millions of years ago. The growth of industry to the level it reached in the 19th century was related to the exploitation of coal, just as the world economy is today related to the extraction of oil from rocks.

RIGHT: Ammonite marble - surface of a specimen containing *Asteroceras* (large shells) and *Promicroceras* (small shells) from the Lower Lias, Somerset, England.

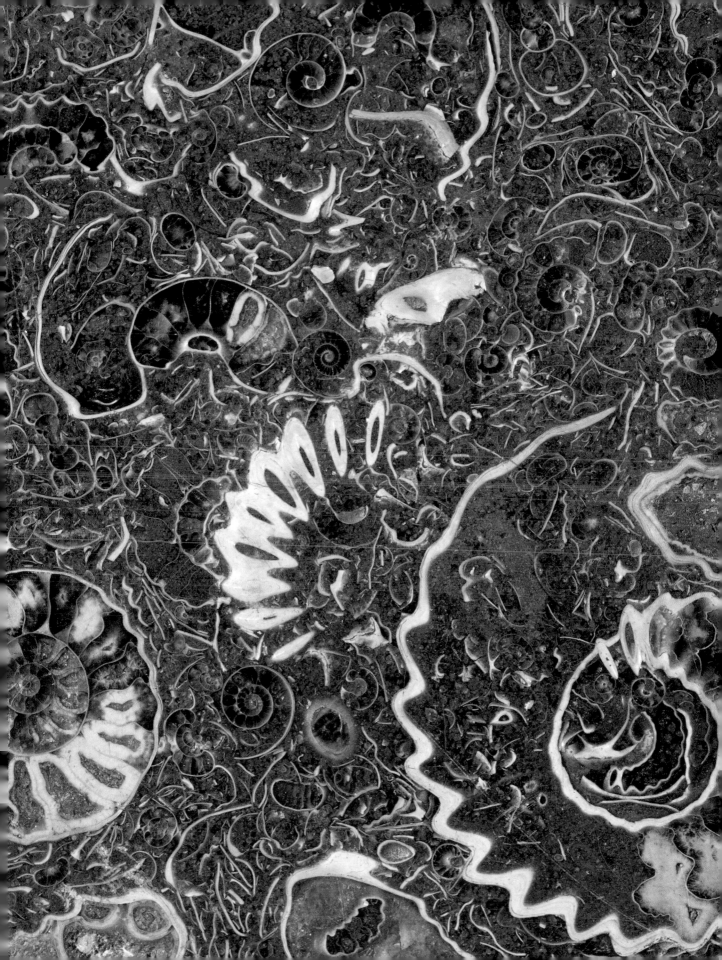

Most of the items that we take for granted, from plastic spoons to television sets, ultimately owe their existence to energy derived from consuming fossil fuels. In times of inflated oil prices this dependence becomes manifest in price rises in all sorts of other commodities that seem at first glance to have nothing to do with oil. It would not be overstating the case to say that Western society owes its present affluence to fossil fuels. As each year passes the resources dwindle alarmingly, and it has become a cliché of modern times that the process cannot go on indefinitely. Humans are exploiting the past, plundering the fossil record, and this can only be done once.

So far we have looked at the ways that fossils have been used to interpret the history of the Earth, and also as items of interest in their own right. It is appropriate now to take a look at the ways in which fossils are of practical use, directly or indirectly. Since the initial impetus that stimulated people to look more closely at fossils was economic (notably the need to produce better geological maps) there has been a constant interplay between the academic side of palaeontology and the ways palaeontological results can be used for industrial means. For such industrial purposes it is neither here nor there to know how trilobites or dinosaurs lived, or to address the problems of classification that exercise the minds of many palaeontologists.

MICROPALAEONTOLOGY

Generally speaking the abundance of fossils is inversely proportional to their size; the tiny ones being the commonest. The chances of finding an identifiable dinosaur or fish from a borehole are very slim indeed, and so the kinds of animals that dominate the landscape in reconstructions of the past do not figure prominently in the records of oil companies. The study of microscopic fossils – micropalaeontology – has become more and more important over the last century. Not only can microscopic fossils be recovered from boreholes with a diameter of a few centimetres, but they can also be teased out of rocks otherwise devoid of organic remains. Some types of microfossil seem to have evolved almost as fast as their larger contemporaries, so they can be used in just the same way, as clocks to measure the passage of geological time. It is their usefulness in stratigraphy that gives them their industrial importance. To correlate between one borehole and the next, or between a whole series of boreholes and those from a different country, one of the simplest and cheapest methods is to identify the fossils. Some of the fossils that are used are introduced in the following sections of this chapter. Micropalaeontologists often become specialists in one group of microfossils of a particular geological period. Now that microfossils have even been found in the later Precambrian rocks, their use spans much of the geological column. Experts on tiny fossils are employed by geological surveys, as well as oil and mineral prospecting companies.

Conodonts

Conodonts are tiny, tooth-like fossils one millimetre or so long, made of calcium phosphate. They can be abundant fossils, occurring in thousands from a kilogramme (2.21 lb) of rock. They are used as important stratigraphic indicators in rocks ranging in age from Cambrian to Triassic, when the conodonts apparently died out. In life, individual conodonts occurred together in clusters consisting of opposing pairs of identical conodonts, and several different

LEFT: Conodonts from different geological periods: (top) Devonian of the Timan-Pechora Russia about 360 million years old (magnification x26); (middle) Silurian of the Ludlow area, Shropshire, England about 420 million years old (magnification x25); (bottom) Ordovician of Estonia, about 465 million years old (magnification x27).

kinds of conodonts often went into these 'apparatuses'. Conodonts are of special use in dating limestones. Because of their phosphatic composition they do not dissolve in acetic acid, so if limestones are put into a bath of acetic acid all the calcium carbonate dissolves leaving behind a residue of insoluble products, including all the conodonts. Conodonts can be hidden in the limestone even if there is no trace of them on the surface. Hundreds of different conodonts have now been described.

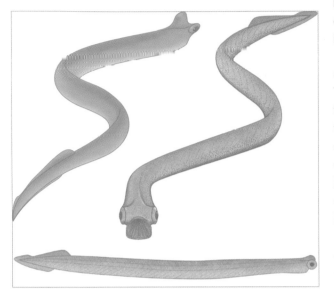

The conodont animal was discovered in the 1980s, and as so often occurs in palaeontology it happened almost by chance. Two palaeontologists were studying Carboniferous arthropod fossils from the Scottish Carboniferous, and conodonts were the last thing on their minds. A curious, worm-like fossil a few centimetres long came to their attention. It was not very conspicuous, which is presumably why it had escaped attention before. Only later did it become apparent that there were conodonts within this creature, and not only did the conodonts belong to a known variety, they were also associated together in a natural apparatus, which made a feeding structure. The chance of this being a fortuitous association was remote. Subsequently, more specimens have come to light, which confirms the correctness of this judgement. The conodont apparatus occupied only a fraction of the length of the whole animal. The animal that bore conodonts was a peculiar one. The long, slim body turned out to be no worm at all, but almost certainly a chordate – a member of the group that includes the vertebrates. The conodont animals were a successful side branch, which apparently left no descendants.

ABOVE: Reconstructions of the conodont animal; note the eyes at the front end. The conodont fossils were the 'teeth' in the front end of the animal.

Ostracodes

It is common to find bedding planes of limestones or shales covered with tiny oval blobs looking like diminutive beans. These are the shells of ostracodes, a group of small crustaceans which are encased in a pair of shells, looking rather like miniature clams. Like all other arthropods they have paired appendages, which they use for feeding and for moving about, and the superficial resemblance to molluscs does not indicate any zoological relationship. Ostracodes are remarkable for having proportionately the largest penis in nature. Most ostracodes are small enough to be classed as microfossils, but a few approach the size of a broad bean, and these species can be as conspicuous as brachiopods when they are broken out of the rock. The smaller species, 1–2 millimetres ($1/25$–$1/12$ inch) across, frequently occur in huge numbers, and under the microscope show a beautiful variety of fine detail, which makes them excellent guide fossils. Ostracodes occur in both marine and freshwater environments

BELOW: Ostracodes: small arthropods with two valves. These examples are from the Jurassic.

(different species of course), and there are specialised species adapted to brackish water conditions as well, so that they are also useful indicators of past sedimentary conditions. The tiny shells are composed of calcium carbonate. The ostracodes have a history possibly extending back to the Cambrian, and they continue as a varied and successful group today. They have been widely used in dating rocks of Mesozoic age, but there seems every reason to suppose that they will be used more widely in dating Palaeozoic rocks as well. The electron microscope has given a new dimension to their study, because it enables even the finest details to be seen at high magnification. Although some ostracodes, particularly freshwater ones, are smooth and featureless, others are covered with pimples and ridges that make them highly distinctive fossils.

Foraminiferans

Most single-celled foraminiferans are very small, less than a millimetre ($1/25$ inch) in diameter, and must therefore be studied under the microscope. Many of the Tertiary species are found in unconsolidated sediments, and can be extracted from the rocks by sieving; the 'forams' can then be picked out by eye under a microscope. These well-preserved fossils are studied using an electron microscope to make high magnification pictures (micrographs). In spite of their small size many foraminifera show a wealth of fine detail; some of them are covered with tiny thorns, and many have long spines, or strange, lip-like structures. Their general shape varies from species that look like tiny necklaces, to others that resemble a small

LEFT: Nummulitic limestone made up of the hard parts of numerous foraminiferans. Shown here at approximately actual size.

RIGHT: *Nummulites gizehensis* embedded in Eocene limestone from 2 km (1¹/₅ miles) northeast of Gizeh, west of Cairo. These specimens have been sectioned naturally by the break in the rock to reveal their internal chamber arrangement (about 7 cm or 2½ in across).

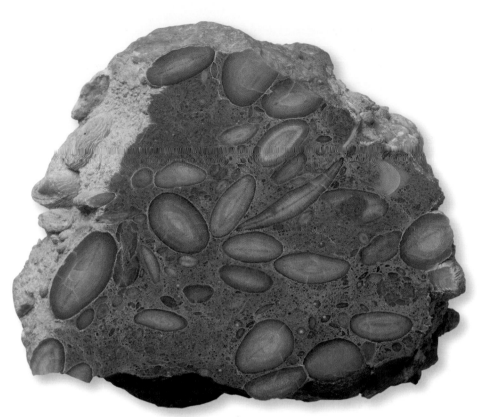

bunch of grapes, and yet others with inflated, globular chambers. The giants among them, nummulites, were abundant in some earlier Tertiary rocks. They resemble small coins in size and shape, but inside are composed of numerous whorled chambers (above).

Forams are a tribute to the flexibility of the single cell, and this flexibility means that they are among the most useful of geological clocks for determining the age of sedimentary rocks. This is especially true of the planktonic species, which are so numerous in the Mesozoic and Tertiary. Oil companies often employ numbers of 'foram men' to correlate the rocks in which oil is found, and some of these experts have become renowned authorities on the history of the group. Since drilling on the deep-sea floor has become a practical possibility, foraminifera have acquired an added importance. Changes in the populations of planktonic species accurately reflect the changes in climate in the more recent geological past, and they can be used to read the ages at which particular pieces of the ocean floor were created at the mid-ocean ridges. Of all fossil animals it is the tiny foraminiferans that have proved most useful in the revolution of geological ideas that accompanied the theory of plate tectonics. Forams have to be used carefully though, because some of the shapes that evolved in the planktonic species were produced independently at different times from different ancestors. Evolution played the same game more than once to produce superficially similar end products. Often it is necessary to look inside the minute chambers to find out the most intimate details of their construction. Many of the older occurrences are in hard limestones from which it is not possible to extract the entire animals, and here thin sections are the only means of studying the evolution of the group. Forams have been used extensively to correlate rocks of Carboniferous and younger age, and really come into their own in the Tertiary, after the disappearance of the ammonites in the Late Cretaceous. Their ubiquity even led one worker to suppose that all rocks (even including igneous ones!) were made of foraminiferans – an impression which might be forgiven after a hard day sorting out hundreds of the animals from residues.

Coccoliths

Even smaller fossils are used to date rocks of Mesozoic and younger ages. Among these, one of the most important groups are coccoliths, minute plates a fraction of the size of a foraminiferan (usually with diameters of only a few microns). In life they were the covers for resting cells of single-celled algae (coccolithophorids), with many coccoliths to a single cell. Some of them are so small that they lie at the limit of resolution for a light microscope, and again their study has been revolutionised by the use of the electron microscope (see p. 59). Coccoliths form beautiful rosettes of calcite plates, every species with the plates stacked in a different fashion. Although so small, they seem to have changed rapidly through time, and are very useful in dating rocks from boreholes and similar small samples. They have their use in criminology, as they can be used to identify the source of even the merest smear of sediment on a suspect's shoe. They can form a large part of the fine fraction of sedimentary rocks, as in the soft, white Cretaceous chalk. Coccolithophorids are very important today in controlling the nutrient balance of the world's oceans.

Microscopic plant remains

The fossils of spores and pollen are extremely small, but they are surprisingly tough. The walls of these tiny grains are composed of a very resistant organic material that serves in life to contain the vital reproductive material, and has a very high chance of becoming fossilised. So insoluble are the walls of pollen grains that they even resist attack by hydrofluoric acid, possibly the most unpleasant and voracious acid there is. The tiny fossils survive after the rocks that contain them are broken down by acids and other chemicals, and it is not uncommon to find these microfossils in rocks

BELOW: Scanning electron microscope image of a coccolithophore, a tiny marine alga made up of just one cell. It has been coloured to show the different features, including the chalky armour of plates (or coccoliths) that surrounds each one.

BOTTOM: Spores from a Middle Devonian lycopod. Although widely dispersed and useful for correlating rocks around the world, their special interest is that they were found in situ in their parent plant. LEFT, viewed from the outside; RIGHT, surface where other spores were attached.

that otherwise lack all trace of organic material. This is because pollen grains are wind-borne, and may come to their final rest in sediments from environments in which other life is lacking. Although small, different spores and pollens have distinct peculiarities that allow the recognition of different species under the microscope. For example, they may be ornamented with ridges or spikes, inflated or compressed, and have many different shapes. They are widely used in dating rocks, and are of prime importance in the dating of freshwater or terrestrial sediments in which the kind of guide fossils that are typical of marine sediments are absent. So important has the study of spores and pollen become that it even has its own scientific discipline – palynology – and palynologists comprise a significant fraction of the palaeontologists employed by oil and mining companies. The identification of spores is vital to the understanding of both the formation and dating of coal and oil deposits. Because the kinds of plants that produce spores and pollen are Devonian and younger in age, the use of these fossils is confined to the later Palaeozoic, Mesozoic and Tertiary.

A particular use for palynology is in the dating of climatic fluctuations during the Pleistocene ice ages. The suite of pollen varieties in these relatively recent rocks can be correlated with species still living, so that cold phases will be associated with a predominance of Arctic species, and warm, interglacial periods will be reflected in the appearance of the pollen of subtropical species of plants. Similar spiky, spherical fossils can also be recovered from marine sediments. These are the cases of resting cells of marine algae (hystrichospheres). They too have become the subject of scientific attention, and palynologists can now contribute much to the dating of marine rocks. It is often possible to recover these tiny fossils from shales that have no other fossils. A related group of fossils, the acritarchs (see Chapter 4) can be recovered from Early Palaeozoic rocks, and even some Precambrian rocks. The future of industrial palaeontology will probably be closely linked with these unassuming, tiny spheres, which can be recovered from the rocks in their thousands.

COAL AND OIL

In 1769 James Watt patented the design of his steam engine, and so began the plunder of fossil fuels. The use of coal as a cheap, accessible source of energy fuelled the growth of industry, and gave industrial nations their ascendancy in world economics which lasts to the present day. The world of trade can be viewed as a vast machine, one which turns on the consumption of energy, mined from the geological record. In the 20th century oil became an even more important source of the energy needed by the machine, which was growing inexorably as population increased, and the material expectations of the workers grew with it.

As the complexity and extent of the industrial process has expanded to its present gross proportions, many people have started to question whether the whole machine might grind to a halt. There will be an end to the reserves of coal and oil; they cannot grow again in the rocks once they have been removed. We have to hope that human ingenuity will be equal to the challenge of finding replacements for fossil fuels. Radioactive minerals and solar energy may go some way towards filling the gap, but new sources of energy will also have to be discovered.

When we burn coal or oil we are releasing the energy that the sun beamed upon the Earth for hundreds of millions of years. All the energy stored in the ground depended on the

photosynthetic activity of plants, whether huge trees in the Carboniferous period or minute, planktonic algae. Coal is the black, carbonised, compressed concentrate of plants, especially trees, and only forms under certain conditions; it is not sufficient to have a stand of trees that gets swamped by sediment. To produce a workable seam of coal generations of trees are needed. Under normal conditions wood decays, and its organic material is consumed by bacteria and fungi until it is friable and porous, and ready to fall back into the soil from where it came. For trees to turn into coal they have to be protected from this normal, aerobic decay. The right conditions for this to happen pertain in swampy environments, where large areas may be fetid with lack of oxygen a few centimetres below the sediment surface. In humid, warm forests like many of those of the Carboniferous, plant growth would have been rapid, but even so it would take several hundred thousand years to accumulate enough material to form a coal seam. It is also essential for the whole area in which the coal is forming to be slowly subsiding, so that in time the tree trunks became part of a thick sequence of rocks.

Over time, the sea would have made periodic advances across the coal swamp, causing a retreat of the trees and burying the potential coal beneath a blanket of marine sediment. At other times increased floods of sediment from rivers would dump their sands and silts. As burial of the coal proceeded, volatile materials were driven off. It requires a great deal of burial (3,000 metres, or nearly 10,000 feet, or so) before the process is advanced enough to produce a coal of much utility. Very deep burial, or the heating effects of nearby igneous activity, are necessary to produce the highest quality coal known as anthracite. All this takes a lot of time, and most of the higher quality coals are correspondingly of Palaeozoic age. Mesozoic or Tertiary coals are often known as lignites – fossil peats in which the process of coalification has not proceeded to the full degree. Carboniferous coals accumulated in a number of separate basins, and the spores of the coal measure plants are useful for correlating rocks from different basins. Some coals are almost entirely composed of masses of such spores. The subsequent history of the coal basins is complicated. They were nearly always fractured by faults, breaking the coal seams up, and so contributing to the dangers of mining which are part of folklore. All coals reveal their plant origin in the impressions of bits of bark, or occasionally leaves, that can be seen on the shiny surfaces. From an environmental point of view, the main problem with coal is that when it is burned it releases carbon dioxide and sulphur products, the former a prime contributor to global warming, the latter a major pollutant.

Oil, the dark fluid that comes gushing from deep wells to stock the refineries of the world, is just as much a product of biological activity as is coal. The ultimate origin of oil is the organic material contained in many sedimentary rocks, including marine mudstones and shales, especially those that accumulated under stagnant conditions. The fixing of the sun's energy in the sea is brought about by photosynthetic plants, particularly single-celled algal plankton. This is the ultimate foundation of the whole economy of the sea, and so the raw material for oil is in a sense the produce of fossil sunshine in the same way as coal. Oilfields have been found in rocks as old as Precambrian. The hydrocarbon compounds typical of oil are released from organic material in sediments by bacterial activity, but at an early stage are dispersed throughout the rock. Time and burial are essential to produce a workable oilfield. During burial and compaction oil is squeezed out of the source rock, together with pore-water, and can then migrate to sites where it becomes concentrated. These are the reservoir rocks, often rocks which would initially have contained no oil at all, but serve to store it because they are highly porous, concentrating all the dispersed hydrocarbons into an economic pool. Finally it is necessary to trap the oil, to prevent it from migrating to the surface. Where oil does reach the surface it produces natural seeps of oil, the most famous of which are asphalt 'lakes' like those in Trinidad, and it is probable that many of the first wells were drilled in the vicinity of such obvious surface shows.

ABOVE: Carboniferous seed-fern plant *Neuropteris*.

The search for oil has led to wider and wider exploration. At first the most accessible fields, like those of Texas, were exploited. The vast resources of the Middle East, which are the main source of supply at the present time, were tapped from somewhat less hospitable territory, but now any area with sedimentary rocks is explored for its oil potential, whether high in the Arctic, or under the sea. The latter areas are naturally confined to the continental shelf, where the sedimentary cover continues to the continental slope – there is no oil in the deep sea. Some of these new areas have proved highly productive, like the British North Sea field, but sooner or later all the possible sites will have been explored, and then there will be nowhere else to drill. The coal industry may anticipate a renaissance before the end of the 21st century.

The classic oil trap is an anticlinal fold in the porous reservoir rock, capped by some impervious strata that make it impossible for the oil to migrate further and escape. The oil is often capped by a field of gas under high pressure. This is why when the drill penetrates the petroleum field the oil gushes out, and if the gas ignites this can result in spectacular, but dangerous 'blows'. Many other kinds of structures can produce important oilfields. The oil can be concentrated into porous, sandy lenticles within otherwise more shaly rocks. In some cases the porous rocks can be fossil reefs, full of corals or bryozoans. Reef rocks of this kind usually have a very high porosity because of the gaps between all the frame-building organisms. They usually also have considerable lateral extent, and for these reasons reef rocks at depth, sealed by less permeable strata, can be very important sources of oil. This explains why hard-nosed oil executives support palaeontological research into the geology of fossil reefs. In the Middle East and elsewhere oil traps have formed against the sides of salt domes. These domes of deeply buried salt are actively rising, almost as if the salt were behaving like an igneous magma. The analogy is not altogether misplaced, because the salt does act in the manner of an igneous intrusion, flowing into the domes and pushing upwards as the load of sediment on either side pushes down. As well as sandstones and reef rocks, magnesian calcite rocks known as dolomites are often full of little cavities that may get filled with oil under the right geological circumstances. This is one rock where the petroleum geologists and the palaeontologists might be forgiven in parting company, because dolomites of this type are one of the poorest kinds of rocks in which to find fossils. All petroleum-bearing rocks, when outcropping at the surface, have a characteristic smell if freshly broken – rather an unpleasant one, reminiscent of greasy rags discovered in a forgotten corner of the garage.

BELOW: The geological circumstances in which one may find accumulations of oil.
TOP, an anticline;
MIDDLE, a reservoir in fossil reef;
BOTTOM, a salt dome trap.

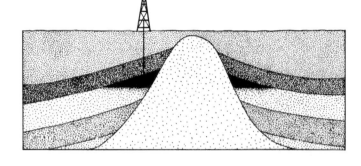

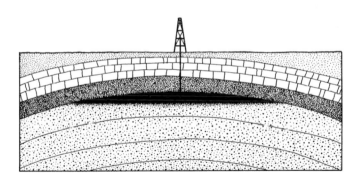

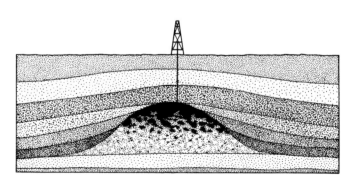

Today, oil is recovered from greater and greater depths, and it is usually impossible to infer whether the right kind of geological structure is present there just by looking at the surface geology. A great deal of exploration is done using geophysical equipment which senses out the most important structures, and records the differences between permeable rocks like sandstones and impermeable shales and mudstones. Drilling now is nothing like the 'wildcat' operation it once was; though even today luck plays an important part in the productivity of any strike. Once drilling starts, the palaeontologists (usually micropalaeontologists) come into their own, identifying the fossils recovered and keeping a log of the age of the rocks through which the drill is passing. In some cases oil is found so consistently in rocks of a particular age it almost looks as if it were seeking out the characteristic fossil. In this circumstance it is likely that the fossil indicates a suitable facies for the formation of an oil reservoir, and here the fossil is used as something more significant than a mere calibration on the age of the borehole rock.

MINERAL DEPOSITS

Many mineral deposits are associated with igneous and metamorphic rocks, and fossils are rarely relevant to their understanding. Other minerals are often found in sedimentary rocks, and fossils are used to correlate the formations containing such minerals in just the same way as in the oil industry. Some of the commercial deposits of iron are in the form of beds of iron-bearing oolites extending over many square kilometres. These include marine deposits with a characteristic fossil fauna consistently associated with the ores. Phosphate deposits, the basis of the fertiliser industry, are also extensively developed in marine sediments. Often such phosphates are found in sites that were once near the edge of a former continent. The phosphate was probably introduced into the sediment as a result of the upwelling of deep oceanic water, which tends to occur in such locations, as it does off Peru today. Upwellings prompt a wide variety of animal life in the seas today, and are also associated with rich fossil faunas in the past. Phosphatic deposits as old as Cambrian have been found associated with a wealth of trilobites. Other peculiar fossil beds have special commercial applications. For example, the diatomaceous 'earth' of the West Indies, a rock almost entirely composed of the tiny fossil shells of diatoms, has a variety of industrial uses, including use as a special filter.

ORNAMENTAL STONES

A walk around a European capital city can often be a synoptic guide to the different fossils to be found in that country, or at least to a range of those to be found in limestones. Particularly in older buildings, when builders had time and money, and labour was cheap, the floors, walls, columns and even the ceilings can be covered with a skin of ornamental stone. Many such facings contain sections of fossils. Many of course, do not, being slabs polished from igneous rocks like granite, or metamorphic rocks like marble. A great variety of limestones have been used as 'marbles' (a confusion of terminology has led all predominantly calcium carbonate facing stones to be called marble, whereas the geologist restricts the term to

BELOW: Fossil ammonites used as an ornamental stone. *Asteroceras marstonense* and *Promicroceras marstonensis*, Marston Magna, Somerset, England.

metamorphosed carbonate rocks) for ornamental purposes, and often the presence of fossils lends them their peculiar charm. Coralline rocks are particularly in demand, and the polished sections afford an excellent way of studying the internal structures of these animals. Other common facing stones include those largely made of the stems of crinoids, which make a bold patchwork of rods and struts in white calcite, contrasting with the darker matrix. In Scandinavia there are beautiful red limestones of Ordovician age containing the remains of straight nautiloids. In southern Europe such ancient limestones are not available, but Cretaceous rocks enclosing the peculiar bivalve molluscs known as rudists are often polished to great effect. Tertiary limestones containing masses of the giant, single-celled foraminifera Nummulites have also been widely used for ornamental purposes, looking like great masses of spiral nebulae adrift in a finely grained groundmass. Limestones formed by layers of algae have been used in the manufacture of ornaments. One such, which was popular in the 19th century, is the Triassic Cotham 'Marble'. This displays patterns looking like a landscape of trees, picked out by the layers of sediment, which are the product of algae and bacteria. The Purbeck 'Marble' was extensively used as an ornamental stone in English cathedrals; columns of this rock surround the main pillars of regular limestone in Salisbury Cathedral. This Jurassic limestone is almost entirely made of the shells of one species of gastropod, and the polished surfaces provide sections through this fossil from every angle. The list could be extended indefinitely: wherever good, homogeneous fossil-bearing limestone is to be found it may be used as a facing stone or as flooring.

DECORATIVE USE OF FOSSILS

Fossils have been used to manufacture many different kinds of decorative objects. They have been found as talismans associated with cave cultures of *Homo sapiens*. The North American Indians used the small oval trilobite *Elrathia kingii* to manufacture a necklace composed of many examples of this fossil. The same species is mined commercially today to produce everything from tiepins to paperweights. The most consistently used fossil material is amber, fossil resin, which was discussed in Chapter 1. Amber fossils include a wide range of insects and spiders. Amber is often shaped into drop-like pieces, and carefully matched for colour, before being mounted in necklaces, pendants and earrings. Some of the 19th century examples of amber jewellery are particularly fine. Fossil wood, where the original living material has been replaced by the hard mineral silica, often retains the finest details of its cellular structure. It takes a very high polish, and is an attractive deep reddish-brown colour when cut. This has been exploited in the manufacture of a variety of table ornaments, and the larger trunks have been cut through to make extremely heavy table tops, which would certainly be immune to the normal stains of domestic use, and must form the oldest tables in existence. In the Jurassic rocks of Yorkshire another kind of fossil wood is preserved as jet – a dense, very dark material that is the origin of the phrase 'jet black'. In the 19th century jet had a considerable vogue in necklaces and the like. The brilliant black colour of polished jet was adopted by Queen Victoria after the death of Prince Albert, and thence by society, as an ornament that was both decorative and decorous.

It has recently become a fashion to treat well-preserved, large fossils as objects of beauty in their own right. Fine examples of ammonites or fossil fish can be found on sale at inflated prices, described as 'Nature's Sculpture'. Good looking fossils can command high prices at

auction, and there is now concern, particularly in the USA, over the commercial exploitation of the biological past. The *Tyrannosaurus rex* specimen known as 'Sue' was sold for millions of dollars. The increased appreciation of the value and beauty of fossils is something to be welcomed. The only problem is that fossils, unlike living animals, cannot breed and replace themselves. Like coal and oil they are a finite resource. Classic fossil localities are easily worked out, and every palaeontologist has had the experience of arriving at a well-known locality to find nothing left but a large hole. Fossils are valuable, each one in its way is a small miracle, but they are most valuable for what they can tell us about the past.

Making a collection

Making collections seems to be almost an instinct with many people. A fossil collection is easy to store and maintain, and is the best way to get to know the many kinds of organisms in their various modes of preservation. Some people may prefer to make collections of fossils from the rocks in the vicinity of their own home, others may deliberately try to get a wide coverage of the different kinds of fossils from rocks of diverse ages.

RIGHT: A variety of echinoderm fossils may form the basis of a specialised collection.

Cidaris coronata
[Goldf.] Schlotheim
Weisser Jura ε

Cidaris coronata
(Schl.)
Form: Weiser Jura ε

Cidaris coronata
(Schl.)
Form: Weiser Jura ε

Cidaris coronata
(Schl.)
Form: Weiser Jura ε

Wright Coll. Purch.
Oct. 1906

Cidaris coronata
(Schl.)
Form: Weiser Jura

Cidaris coronata
(Schl.)
Form: Weiser Jura ε

Cidaris coronata
[Faltd.] Schlotheim
Form: Weisser Jura
Loc: Randen
Brit. Mus. Geol. Dept. E 2723

Cidaris coronata
[alveoli] Schlotheim
Form: Weisser Jura ε
Loc: Randen
Brit. Mus. Geol. Dept. E 2721

Cidaris coronata
(Schlott.)
Form: Weiser Jura

Cidaris coronata (Schl.)
Form: Weiser Jura ε
Nattheim

Cidaris coronata (Schl.)
Form: Weiser Jura ε

(alveoli)
Cidaris coronata
(Schloth.)
Form: Weisser Jura

Cidaris coronata
(Goins?)
Form: Cornaliae Stappen

Calvados
July 1857 Coll. bought
57730.57 Re-reg'd from
E 39319-20
Brit. Mus. Geol. Dept. E 4730

Cidaris
coronata
Form:

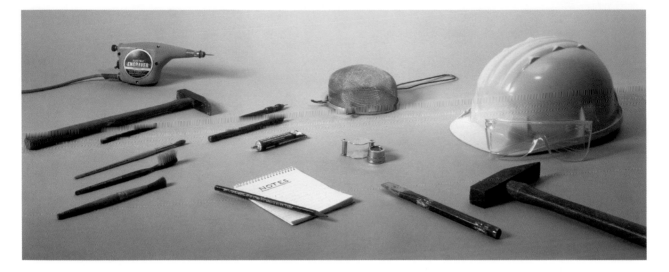

ABOVE: Implements for use in the field and for cleaning fossils: tools for extracting, brushes for cleaning, sieve for extracting from clay, hand lens, and helmet and goggles for protection.

It is surprising how quickly a collection grows; if given a chance it soon begins to oust the collector from house and home. This chapter gives a little practical advice on how to find fossils, clean them, identify them and store them. There is no doubt that some people have an almost supernatural facility for finding fossils. Fossils seem to fall out of the rock for them, whereas others labour long and hard with no reward. There is no substitute for returning time and again to the same site and carefully examining anything that looks organic. The only two essential tools for this are a good geological hammer and a hand lens. Geological hammers come in two main varieties: those with a wooden shaft, and those in which the head is welded to a hardened steel shaft. Either is perfectly satisfactory for hammering sedimentary rocks. Otherwise all that is needed are old newspapers (for wrapping) and patience. It is good practice to label each specimen in the field. It is amazing how your memory lets you down.

WHERE TO FIND FOSSILS

There are no special rules about where to find fossils. There are certain restrictions: if rocks have been heavily metamorphosed there is a slim chance of recovering good fossils, but even this is not impossible, and there are examples known where fossils have survived the most appalling maltreatment by heat and pressure. In general, unaltered sedimentary rocks are fossil-bearing. There are a large number of known localities, that is, places where fossils have been recovered for a long time, and which are recorded in the geological and popular literature. At such places fossils are sure to be found, but it has to be remembered that generations of hunters have been there before, and that the inspiring specimens now residing in museums were probably collected 100 years before, when the ground was in prime condition. Nowadays the best collections can often be made from the same geological horizon as the classic localities, but from sites a few hundred metres to a kilometre away. There are some areas within which the rocks are thoroughly saturated with fossils; in these cases it is just a question of having enough time to gather all that may be collectable. The Jurassic rocks of the Dorset coast in southern England are one of the best-known rock groups of this kind. Formation succeeds formation, all of them rich in fossils. Even if the most spectacular fossils have been removed there is no chance of the localities being 'collected out'. The same might be said of Cretaceous deposits in large areas of the interior of the USA, or the Devonian of Canada. Such areas are certainly the most encouraging in which to make a first collection. They usually represent former deposition in shallow marine habitats, rich in life, where the sediments have been little disturbed subsequently, and where the soft sedimentary rocks yield up their treasures with little resistance. Even in these well-known localities a lucky hammer blow may turn up a species new to science.

Such prolific sites apart, most sedimentary rocks do yield fossils with prolonged searching. Sandstones are usually the least rewarding. Some sandstones were deposited in desert environments and scarcity of fossils in these is hardly surprising. Others are turbidites, and fossils are rare in these too, although the presence of trace fossils should not be ignored. Limestones only occasionally lack any organic remains (apart from Precambrian limestones). If the reader is attracted by finding older fossils there is the increased possibility of structural complication in sedimentary rocks of this antiquity. Cambrian to Devonian rocks (and younger rocks in some areas) are often affected by cleavage, which make the rocks split in directions unrelated to the original bedding (in which the fossils lie). Even so, it is quite feasible to recover fossils by looking for traces of the true bedding and smashing the rock so that it breaks in approximately the right direction. This can be a heart-breaking process, trying the patience of even a professional. If the rocks are folded as well as cleaved, it is possible to find places where the cleavage and bedding coincide, and this is the premium site in which to search for fossils.

Many rock formations, although they do yield fossils, have them concentrated into small pockets in certain localities. This particularly applies to formations of freshwater or terrestrial origin, probably reflecting the patchy distribution of lakes and pools where fossils can be preserved. The Old Red Sandstone formations of Wales and Scotland locally contain fish, eurypterids and other exciting fossils, although great stretches of the rocks can be disheartening to the casual searcher. Here persistence, and a little luck, are indispensable, and there is always a thrill in the discovery of any well-preserved fossil.

Coal measures usually yield fine remains of plants, but the best of these are not in coal itself, but in associated shales. There is always the possibility of finding one of the seams in which insects, like giant dragonflies, or vertebrates are also preserved. Rarer fossils are often concentrated in this way into particular bedding planes, and so are easily worked out by the over-zealous collector. It is essential to note the exact horizon of any unusual find, so that it can be located again.

There are many kinds of exposure in which it is possible to find fossils. The most obvious exposures are those in sea cliffs, and these also provide the best sections through different formations. In desert regions exposure can also be good, and the same is true of Arctic areas or high mountains, although collecting there is not without its attendant hazards. Never take chances climbing high or crumbling rocks on the supposition that the best fossils are to be found slightly out of reach! In domestic landscapes like that of England or the eastern USA most of the best inland exposures are in quarries or on road cuttings. It is always worth examining temporary exposures cut during road widening. Here, local knowledge is a great advantage: if you are 'on the spot' you can collect quickly before the site is backfilled and the chance lost forever. Occasionally these temporary cuts open up seams of important fossils that have never been found before. It is always necessary to ask permission to enter any working quarry; most quarry owners are quite happy to let in the foraging palaeontologist, but they do not wish to have accidents happening on their property.

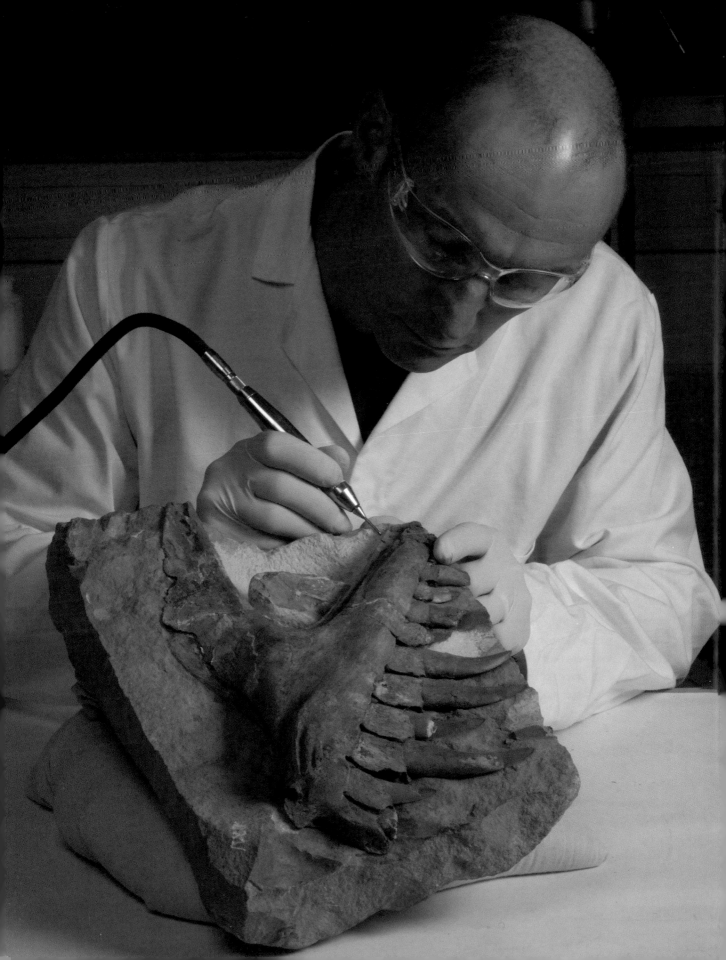

together with any peculiarity of particular beds, like the presence of trace fossils, or a change in colour. Such observations are the basis of stratigraphy, and a well-kept notebook will jog the memory when the fossils are being examined back home.

OPPOSITE: Maxilla (cheekbone) from a *Megalosaurus hesperis* – a theropod dinosaur – from the Middle Jurassic of Dorset, England.

CLEANING FOSSILS

There is always a temptation to chip out a sea urchin or a trilobite as completely as possible while still in the field, just to get a proper look at it. This is nearly always a mistake: a careless blow with a hammer can destroy what might, with patient cleaning, turn out to be a fine specimen. It is far better to wrap the specimen up carefully for cleaning at home. It is important to keep the other half (counterpart) of the specimens, which often show details lacking on the specimen itself. Most fossils are not so fragile as to need special treatment in the field; careful wrapping is sufficient. Some, however, are very delicate and need to be toughened up before removal by using solutions of resins. If one is lucky enough to stumble across the remains of a large vertebrate, the best thing to do is to leave it exactly where it is, and contact a museum with the expertise to extract it properly. Otherwise, it is possible to destroy vital information.

Once the fossils are safely home and unwrapped, the cleaning process can begin. Some fossils tend to crack out completely, graptolites in shales for example, and little needs to be done to these. Sometimes, however, the shale still partly covers the specimen, and in this case a sharp tap with a small chisel usually suffices to break the shale along the bedding plane containing the rest of the fossil. The crucial point to remember when trying to dig out a fossil is that damage will result if the cleaning implement is much harder than the fossil and used directly against it. Fossils in shales and mudstones tend to be soft, and great care is necessary to avoid damaging them. A needle mounted in a pin vice can be used to gently remove any covering shale. This should be done by pressing obliquely on to the covering rock, and not by stabbing at the fossil itself, which nearly always results in unsightly pinpricks.

In some cases the enclosing matrix is softer than the fossil. This is the easiest case to deal with, because the enclosing rocks can usually be removed by scrubbing. Fossils from Cretaceous chalk can be cleaned using a small bristle (but not wire) brush with water. Cleaning is much more difficult if the matrix is harder, or about the same hardness as the fossil material. There is usually no choice here but to clean off the matrix gradually by hand (relying on the tendency for rock to break off around the fossil rather than through it), at the interface between the fossil and the enclosing rock. A mounted needle can be used for this (the best are the needles used for 78 rpm records, which can still be bought in junk shops), and there are various manufactured appliances that allow the needle to vibrate very fast, producing an instrument that chips away the matrix more rapidly. Cleaning fossils in this way is quite a skilled operation, and if this is being tried for the first time it is best to start on one of the more unimpressive specimens, however tempting it might be to start with the prize exhibit. Sometimes nature lends a hand with the process, and natural weathering may have already etched out a slightly harder fossil from the surrounding rock. Many of the most impressive museum specimens have been prepared naturally for exhibition in this way, but the majority of these sites have since been picked bare of such treasures, and nowadays there is usually no substitute for hard work.

In sandstone preservation particularly, but also in some limestones, the actual shell of the fossil has often been dissolved away, and what remains are internal and external moulds. The internal mould usually comes away with little trouble. The external mould should be washed clean of any dirt, which may have to be removed by gently rubbing with a toothbrush. A perfect replica of the exterior surface of the fossil may now be made by taking a cast from the external mould. Various preparations are suitable for this purpose: a plasticine squeeze can give a quick impression, but does not take up the finest detail. For this a latex rubber solution should be used, or some of the modern resin preparations, the latter having the advantage that the casts are more or less permanent. In fact, casts of this kind are every bit as good, and in some cases better, than having the fossil itself, particularly since the internal mould also gives you all the details of the internal surfaces of the fossil.

In some limestones the fossils have been silicified (replaced by silica) while the matrix consists of calcium carbonate. In this case the rock can be dissolved in dilute acetic or hydrochloric acid (with care!), and the fossils will be left behind. After repeated washing and drying a collection of fossils preserved in this way can be mounted on slides, or in small boxes. Any phosphate fossils (such as lingulate brachiopods) are left untouched when limestones are dissolved in acetic acid, and some beautiful and surprising fossils can be recovered in this way, even when there is not much to show on the surface of the rock.

IDENTIFYING FOSSILS

While it is satisfying to make a collection of beautiful specimens of fossils for their own sake, it becomes even more so if the specimens are identified and classified. Many people find that a general collection of fossils soon begins to take up too much room, and that they prefer to specialise in one kind of fossil (such as trilobites) or in those from a particular area or age. Putting a name to a fossil may seem a rather arid exercise, and so it is by itself. But the real point is that it is not possible to identify correctly a fossil without looking at it very carefully, appreciating the fine points of its construction, and becoming familiar with a whole range of related animals or plants. Therefore to identify is partly to understand.

To identify the general kind (e.g. phylum) of animal or plant is usually a simple task. With experience it is also easy to identify the fossil to within narrow limits (i.e. to order level). There are only a relatively few major kinds of brachiopods, for example, and their general features can be mastered with a little experience. The problems start when a precise identification is required, to genus or even species. There is no easy way to do this, and sometimes even the expert in the group of fossils will have problems. In some cases a species identification may not even be possible: for example, many brachiopods are identified from their internal structures, so if your fossil does not show these it cannot be specifically identified. It is vital to know the precise formation and locality from which the fossils were recovered. Fortunately there are reference collections in museums, which have fine specimens on display that have (one hopes) been identified by an expert. These are often arranged rock formation by rock formation, and so it is possible to home in on a series of species with which the one in hand is to be compared. It usually happens that your specimen is not exactly the same as the ones on display; the best name is then that of the closest matching species. Most fossil animal populations, like many living ones, include a certain amount of variation, so that it is unlikely that any two fossils will be precisely the same, even

if they belong to a single species. Among palaeontologists there are often arguments about whether or not a population of fossils slightly different from other examples of a species is really a different species or just a local variant.

The next best thing to using a comparative collection to identify a fossil is to use books with pictures of fossils as the basis for identification. Any available books only give a selection of typical fossils from a region, but this is a good start, and because most books will show the most common fossils you are likely to have at least some of your finds illustrated. Often, a fossil is found which cannot be closely matched with any of the species illustrated in these books. In this case the best approach, if the fossil is an invertebrate, is to obtain the relevant volume of the *Treatise on Invertebrate Palaeontology*. This rather intimidating work is a compilation of all the different kinds of fossils known at the time it was published (different years for different volumes); each volume concerns one type of fossil (e.g. brachiopods or trilobites), and contains descriptions and figures of all the different genera of the animals. It can take a while to learn how to use one of these volumes, and the pictures vary in quality. This only enables an identification of the genus of the fossil, but this is adequate for most purposes. The volumes of the *Treatise* are not likely to be stored in a local library, but they can be ordered through libraries, and they are certain to be stocked in university libraries. Finally, many museums offer an identification service. Occasionally this kind of enquiry will result in the discovery of a new species, and then the finder may be asked to donate it so that it can be subjected to proper scientific scrutiny. This underlines how important it is to record the precise details of where a fossil is found. New discoveries are quite often made by amateur collectors, particularly when they return again and again to a favourite site to find the rarer fossils in the fauna. One of the greatest excitements of palaeontology is the possibility that something entirely new will turn up; any hammer blow could be the one that breaks out the new discovery. It is this that sustains the searcher through long hours where nothing at all is discovered.

STORING A COLLECTION

As the man who made a fortune from selling 'pet rocks' in the USA shrewdly realised, fossil specimens are remarkably easy to look after. Most fossils do not deteriorate with time, and all they require is room for storage in a dry place. A few fossils look superficially attractive if they are varnished; this is not a good practice, however, because the varnish obscures a lot of finer detail, and is very difficult to remove once it dries. All fossil specimens should be labelled either with a direct identification and locality, or with some sort of code that refers you to a catalogue. When a collection begins to get large, it is impossible to remember all the details about when and where a fossil was collected, and how it was identified.

Some kinds of fossils are prone to decay. This is particularly true of those preserved in iron pyrites, such as ammonites and Tertiary fruits and seeds. They acquire feathery growths of crystals as the pyrite begins to react with the atmosphere. The eventual outcome is that the fossil collapses into a heap of dust. Varnish does not greatly inhibit the process of decay. There are inert fluids on a silicone base in which fossils like these can be stored. A cheaper

BELOW: Collection of ammonite fossils properly labelled and stored in boxes in a drawer.

alternative is glycerine, but this is messy, and gradually takes up water from the atmosphere, so has to be stored in airtight jars. How the fossils are arranged is a matter of taste. Some people prefer to order them phylum by phylum, others according to their geological age. If the intention is to make a very detailed collection from one area or formation, the arrangement might be by locality. In many ways the last kind of collection is the most satisfying, because it does not take long before the commoner fossils become old friends, and can be used to trace a single geological horizon from one locality to the next. With prolonged collecting, specimens as good as any found in museums will turn up, and the collector will begin to know the faunas as well as any expert. Geological time is immensely long, and the volume of fossil-bearing rock surpasses computation. There is plenty of scope for those with time and patience to make a real contribution to the knowledge of the history of life. The evidence is just waiting to be collected.

Further information

FURTHER READING

African Exodus: the origins of modern humanity, C. Stringer and R. McKie. Cape, London, 1996.

At the Water's Edge, C. Zimmer. Simon & Schuster, 1999.

The Complete World of Human Evolution, C. Stringer and P. Andrews. Thames & Hudson, London, 2005.

Dinosaur. Eyewitness Guides 13, D. Norman and A. Milner. Dorling Kindersley, London, 1989.

The Dinosaur Hunters, D. Cadbury. Fourth Estate, London, 2000.

Echoes of Life: what fossil molecules reveal about earth history, S. Gaines, G. Eglinton and J. Rullkötter. Oxford University Press, New York, 2008.

The Ecology of Fossils: an illustrated guide, W. McKerrow (ed.). Duckworth, London, 1978.

The Eternal Frontier: an ecological history of North America, T. Flannery. Heinemann, London, 2001.

Evolution, C. Zimmer. Heinemann, Oxford, 2002.

The Evolution Revolution, K. McNamara and J. Long. John Wiley & Sons, London, 1998.

Evolution of Fossil Ecosystems, P.A. Selden and J. Nudds. Manson, London, 2004.

Extinction: how life on Earth nearly ended 250 million years ago, D.H. Erwin. Princeton University Press, 2006.

The First Fossil Hunters, A. Mayor and P. Dodson. Princeton University Press, 2001.

Fossil. Eyewitness Guides 19, P. Taylor. Dorling Kindersley, London, 1989.

From the Beginning, K. Edwards and B. Rosen. Natural History Museum, London, 2000.

Graptolites: writing in the rocks, D. Palmer and B. Rickards (eds.). Boydell Press, Suffolk, 1991.

The Greening of Gondwana, M.E. White. Reed Books, 1986.

Homo Britannicus: the incredible story of human life in Britain, C. Stringer. Allen Lane, London, 2006.

In Search of the Neanderthals, C. Stringer and C. Gamble. Thames & Hudson, London, 1994.

The Illustrated Encyclopedia of Dinosaurs and Pterosaurs: an original and compelling insight into life in the dinosaur kingdom, D. Norman and P. Wellnhofer. Salamander Books, London, 2000.

The Legacy of the Mastodon: the golden age of fossils in America, K.S. Thomson. Yale University Press, 2008.

Life: an unauthorized biography. A natural history of the first four thousand million years of life on Earth, R. Fortey. Harper Collins, London, 1997.

Life on a Young Planet: the first three billion years of evolution on earth, A.H. Knoll. Princeton University Press, 2003.

The Map that Changed the World: the tale of William Smith and the birth of a science, S. Winchester. Viking, London, 2001.

The Meaning of Fossils, M. Rudwick. University of Chicago Press, 1972.

National Audubon Society Field Guide to North American Fossils, I. Thompson. Knopf, 1982.

The Natural History Museum Book of Dinosaurs, 2nd edn., T. Gardom and A. Milner. Carlton Books, London, 2001.

The Rise of Animals: evolution and diversification of the kingdom animalia, M.A. Fedonkin, J.G. Gehling, K. Grey, G.M. Narbonne and P. Vickers-Rich. John Hopkins University Press, Baltimore, 2007.

Scenes from Deep Time, M. Rudwick. University of Chicago Press, 1992.

Snowball Earth: the story of the great global catastrophe that spawned life as we know it, G. Walker. Bloomsbury, 2003.

Trilobite! Eyewitness to Evolution, R. Fortey. HarperCollins, London, 2000.

Trilobites, H. Whittington. Boydell Press, Suffolk, 1992.

The Triumph of Evolution, N. Eldredge. W.H. Freeman, London, 2000.

Your Inner Fish: a journey into the 3.5-billion-year history of the human body, N. Shubin. Pantheon Books, New York, 2008.

WEBSITES

NB. Website addresses are subject to change.

Australia Online Museum
www.amonline.net.au/eanh sciences/palaeontology.htm

Buena Vista Museum of Natural History, Kern County, California
www.sharktoothhill com/

Carleton University, Canada
http://hoopermuseum.earthsci.carleton.ca/hvpmdoor.html

The Field Museum of Natural History
www.fieldmuseum.org/

Museo Nationale de Ciencias Naturales, Madrid
www.museociencias.com/english/index1.html

Museum of Paleontology, University of Michigan
www.ummp.lsa.umich.edu/index1.html

National Museum of Natural History, Smithsonian Institution
www.nmnh.si.edu/paleo/

Natural History Museum, London
www.nhm.ac.uk

The Palaeontological Association
www.paleosoc.org/

PaleoNet
www.nhm.ac.uk/hosted_sites/paleonet/PalAss/index.html

Paleontological Research Institution
http://www.priweb.org/

Pella Museum, Jordan
www.pellamuseum.org/

Plant Palaeobiology Group, Royal Holloway, University of London
www.gl.rhbnc.ac.uk/palaeo/palaeo.html

The Royal Tyrrell Museum
www.tyrrellmuseum.com/

Glossary

Abyssal At great depth, off the edge of the continental shelf.

Appendages The limbs, gills and antennae of arthropods.

Aragonite Form of calcium carbonate employed in the construction of shells by some marine animals.

Astogeny Growth of a colonial organism by addition of individuals of the colony.

Bedding plane Plane parallel to the former sea floor (or freshwater equivalent)

Benthic (or benthonic) Describing organisms that live (or lived) on the sea bottom.

Calcite The common mineral form of calcium carbonate, of which many fossils are made.

Chert A hard sedimentary rock composed of fine–grained silica.

Cleavage Tendency for metamorphic rocks (e.g. slates) to break at an angle to the bedding plane (hence, plane of cleavage).

Conglomerate Coarse, pebbly sedimentary rock, such as the deposit of an ancient beach.

Correlation The process of establishing the time–equivalence (or otherwise) of sequences of rocks.

Epicontinental Surrounding the edges of continents, especially of shallow seas.

Era Major division of geological time: Proterozoic, Paleozoic, Mesozoic and Cenozoic (Tertiary and Quaternary).

Exoskeleton The exterior skeleton of the arthropods.

Facies Rock type (or collection of rock types) representing a particular environment where the rocks were deposited (e.g. reef facies, lagoonal fades); also applied to faunas reflecting the same environment.

Fauna An assemblage of fossil animals from one site or age (a flora is the botanical equivalent).

Filter–feeder Animal that lives by filtering out small particles (usually plankton) from the sea, or fresh water,

Helical coiling The upward–spiral kind of coiling typical of many snails and a few ammonites.

Homeomorph An animal that resembles another, possibly because of a similar mode of life, but is not really biologically related.

Igneous Rocks that have formed from the cooling of hot, liquid magma, ultimately derived from deep in the Earth, and of course without fossils.

Internal mould Natural cast in sediment of the inside of a shell or other fossil.

Intrusion Mass of igneous rock, which intrudes into the surrounding strata.

Living fossils Term applied to animals or plants that have survived for a long time, or at least from a time when there were many more of their kind.

Marine transgression Invasion of the sea over land area, caused by relative rise in sea level.

Metamorphic Rocks that have been altered by heat and/or pressure, usually because of deep burial or involvement in mountain–building episodes.

Oolite Sedimentary rock (usually limestone) composed of spherical ooliths, and usually formed in shallow water,

Pelagic Free swimming in the oceans.

Phyla (singular, phylum) Major zoological unit of classification, indicating broadly related animal groups.

Planktonic (or planktic) Passively floating in the oceans (noun, plankton).

Radiometric age Age given in years using the natural 'clock' of radioactive minerals.

Regression Draining of the sea from a continental area; the opposite of transgression.

Shield areas Large areas composed of Precambrian rocks that have acted as coherent, stable blocks for hundreds of millions of years.

Silica One of the most abundant compounds in nature, silicon dioxide, of which quartz and chalcedony are two of the commonest forms, and which is used as a skeletal material by a few organisms.

Stratigraphy That part of geology concerned with the description of the relationships of rocks in the field, and their correlation.

Subduction zone Area where oceanic crust is plunging downwards beneath an adjacent continental block

Suture line Line marking the junction of the chamber wall with the shell in ammonites and nautiloids.

Test The 'shells' of echinoderms or foraminiferans.

Turbidite Sedimentary rock formed by the action of a turbidity current.

Unconformity Break between two sequences of rocks; the lower sequence is often tilted, uplifted and eroded before the deposition of the overlying one (angular unconformity).

Zone Basic unit for correlation, a unit usually typified by a characteristic assemblage of fossils belonging to one or more groups of organisms.

Zooid Individual animal in a colonial animal – applied particularly to bryozoans and graptolites.

Index

Picture credits

p.19 bottom © Professor Wilhelm Sturmer; p.22 © Angela Milner; p.24 Reproduced by permission of the British Geological Survey. © NERC 2008. All rights reserved; p.31 © Martin Bond/Science Photo Library; p.32 Mercer Design © The Natural History Museum; p.34 Mercer Design © The Natural History Museum; p.36 bottom, p.38 top © Perks Willis Design; p.39 © Istockphoto; p.41 top © Perks Willis Design; p.41 bottom © Francois Gohier/Ardea; p.42 top Mercer Design © The Natural History Museum; p.44 © Brian Rosen; p.52 top © John Chapman; p.52 bottom © Perks Willis Design; p.53 © Linda Pitkin; p.59 top left © Markus Geisen/ NHMPL; p.67 Mercer Design © The Natural History Museum; p.87 left © Roger Steene/ imagequestmarine.com; pp.142–3 © John Sibbick/NHMPL; pp.145–146 © Ray Burrows/© NHMPL; p.148 top © Dr Andrew Milner; p.150 © Geological Museum of China/NHMPL;

pp.151–153 Ray Burrows/© NHMPL; p.155 © Ray Burrows/© NHMPL; p.156 top © Lars Ramsköld; p.156 bottom Ray Burrows/© NHMPL; p.157 © Derek Briggs; pp.158–9 © Georgette Douwma/cience Photo Library; p.160 © NASA; p.162 © Karl Stetter; p.163 © NASA Jet Propulsion Laboratory (NASA–JPL); p.164 © Bill Schopf; p.165 bottom © Brian Rosen; p.166 © Pamela Reid; p.167 © Bill Schopf; p.168 © N Butterfield; p170 left Ray Burrows/© NHMPL; p.170 bottom © Richard Fortey; pp.172 & 173 bottom right Ray Burrows/© NHMPL; pp.178–9 © De Agostini/NHMPL; p.182 MercerDesign © The Natural History Museum; pp.183 & 184 left Ray Burrows/© NHMPL; p.188 bottom © De Agostini/NHMPL; p.190 Angela Milner/ MercerDesign © The Natural History Museum; p.193 Mercer Design/Ray Burrows/© NHMPL; p.197 © Donald C. Johanson, Institute of Human Origins; pp.198 & 199 © John Reader/

Science Photo Library; p.203 bottom © Georgian National Museum. Photo: Guram Tsibakhashvili; p.204 © John Sibbick/NHMPL; p.206 © Laurent Orluc/Eurelios/Science Photo Library; p.207 top, bottom Lisa Wilson © The Natural History Museum; p.208 © Vanhaeren & d'Errico; p.212 © Peter Snowball/NHMPL; pp.216–7 © Tkachev Andrei/Photoshot; p.220 Mercer Design © The Natural History Museum; p.221 © Ted Daeschler/Academy of Natural Sciences/VIREO; p.226 top © Mark Purnell; p.229 top © Markus Geisen/NHMPL; p.233 © Ray Burrows/© NHM; p.234 © Dan Bosence. All other images © The Natural History Museum, London.

Every effort has been made to contact and accurately credit all copyright holders. If we have been unsuccessful, we apologise and welcome correction for future editions and reprints.